Technology and Workflows for Multiple Channel Content Distribution

Technology and Workflows for Multiple Channel Content Distribution

Infrastructure Implementation Strategies for Converged Production

Philip J. Cianci

Focal Press
Taylor & Francis Group

NEW YORK AND LONDON

First published 2009
This edition published 2013
by Focal Press
70 Blanchard Road, Suite 402, Burlington, MA 01803

Published in the UK
by Focal Press
2 Park Square, Milton Park, Abingdon, Oxon OX14 4RN

Focal Press is an imprint of the Taylor & Francis Group, an informa business

Practitioners and researchers must always rely on their own experience and
knowledge in evaluating and using any information, methods, compounds, or
experiments described herein. In using such information or methods they should
be mindful of their own safety and the safety of others, including parties for
whom they have a professional responsibility.

Product or corporate names may be trademarks or registered trademarks, and are
used only for identification and explanation without intent to infringe.

Library of Congress Cataloging-in-Publication Data
Application submitted.

British Library Cataloguing-in-Publication Data
A catalogue record for this book is available from the British Library.

ISBN: 978-0-240-81172-7 (pbk)

Table of Contents

Acknowledgments

Writing a book like this forces one to draw from a lifetime of experiences; things I did decades ago have come to have relevance now.

If Wayne Carpenter and I had stayed in the "Big D," I'd be rich, famous, and dead by now! Those bands we formed and managed in our teens were amazing. From the Gent's and Ted Mack's Amateur Hour to Chuck McCann and auditioning on the Soupy Sales set, the "last night" at Arthur's disco in 1969, and then Woodstock and Boston ... the bond between our spirits can never be broken.

Returning from the Woodstock Festival in 1969 and needing a job, I tried a local business in my home town of Harrison, NY and got my first real job. It was at color media innovators Rosco Labs. I am indebted to Stan Miller and the late Lenny Kraft for breaking me into theatrical production. Their enthusiasm for their products was infectious.

Without Bob Regina, who hired me at ESPN and gave me the opportunity to explore and learn about how a TV superpower creates content and conducts business, I never would have moved from broadcast engineering to the bigger picture, encompassing not just the technology but the creative and business factors relevant in the media business. The department he put together (BITS then BMTG), a group that addressed IT in broadcast, was a source of esoteric IT wisdom. I learned just how complicated networks, storage, and computer platforms can be, as they educated vendors and system integrators in order to get ESPN's Digital Center up and running.

Steve Keaney at ESPN Creative Services clued me in to the mix between creative and technology in a broadcast environment. My conversations with Steve were always on the leading edge of what might be doable. I certainly learned a lot about creative application of technology.

Then there were my colleagues at Philips Research—Kees van Zon, Dave Bryan, Johan Johnson—and Michael Gerrings, Gert Gertz, and Jack Koopers at the Nat Lab in Eindhoven. Their support and guidance in establishing a Video Simulation Infrastructure, beginning in 1999, provided the foundation for my segue from DTV R&D to broadcast facility systems engineering.

Thanks to all the pros at Focal Press for providing the mechanism to create this book; especially, Angelina Ward—who saw this book proposal through an aborted review, a failed review, and finally a successful review—and Carlin Reagan, who worked as the in-house development editor and saw it into production. Thanks also to assistant editor Kara Race-Moore and project manager Phil Bugeau. And thanks to Eric Schumacher-Rasmussen, editor of *Streaming Media* and my development editor, who tirelessly worked with me through the crunch to get this book published in time for NAB 2009. Without his professionalism and perseverance, I would have given up all hope.

I owe many thanks to Brad Dick, editor-in-chief of *Broadcast Engineering* magazine, for his longstanding support. While I was editor of the e-newsletter *Transition to Digital*, I developed many topics that are discussed in this book. I thank Penton Publishing for their permission to use and adapt some of the content that appeared online and now appears in various parts of this work.

My thanks go to James Snyder colleague and friend who reviewed Chapter 7. Hopefully, I incorporated his corrections accurately and completely; if not, the fault is entirely mine.

My appreciation is also extended to Communication Engineering, Inc. for their support in writing this book; a great bunch of people and a collegial work environment.

And of course, my deepest gratitude to my sweetheart Marcie, who gave me just the right jolt to get me to finish the last 5%, always the hardest part.

For my Mother

Rose

Who will always view life through the eyes of a native New Yorker

Introduction

Convergence is a term that has gone in and out of favor in the media business. Many people in the media and technology worlds, from Emperor Gates to Steve Case, have envisioned a world where all content was consumed on a single do-everything device available over a single delivery channel. One faction saw the PC as the only device anyone would ever need for entertainment, work, and communication. Naturally, the Internet would be the only delivery channel necessary. Such was conventional wisdom in 1995.

That dream of realizing a PC-centric media world came face to face with the real world, and the meeting wasn't pretty. The broadcasting industry wasn't about letting the PC take the place of good ol' TV. The 1995 EIA/IEEE Conference in Philadelphia described a three-device scenario. In this world, consumers, in a lean-back mode on their living room couch, would always want their entertainment supersized, with the highest possible audio and video quality. After all, this was, at least in part, the motivation behind the recently proposed DTV terrestrial transmission standard that included HDTV and 5.1 surround sound—to bring a theater-like experience to the living room, a communal event.

PCs were thought of as lean-forward tools for work. Remember, in 1995 the Internet was not the ubiquitous consumer "appliance" that it is today. Although it was growing, it was still primarily a communications link between government, corporate R&D, and educational entities. Besides, network bandwidth was miniscule and PC modems operated at 9600 baud. Transfers of large file (1 MB) were time-consuming, and audio and video compression codecs were not adapted to the consumer domain. CRT displays were 15 inches. PCs were certainly one-person consumption devices.

All of these devices ran on house power, 110V AC provided from the power grid on its own wiring to a residence or business. Enlightened, worldly minds, even pre-9/11, realized then the need for lifeline, emergency communication capabilities. Hence, the third device category would be a telephone. The telco infrastructure of landlines provided power independently to phones and operated even if all AC power was lost.

Time and technology marched on, and what was cutting edge and visionary in 1995 gave way to the realities of the marketplace and to emergent technologies. PCs got faster and memory got cheaper; the Internet became a consumer phenomena. Cable TV challenged and then surpassed traditional over-the-air audience numbers. Cellular phones went from a novelty that only the rich could afford to a necessity for all, replacing the landline as the third element in the triumvirate of devices.

Yet through all this change, the three-screen model still remains valid today.

SOME THINGS DO CHANGE

Media consumption practices and habits have evolved. A frequent mantra at media industry conferences touts how today's media consumers want their content on any device, at any time, and in every place. Watching the Super Bowl on my cell phone may not be my first choice, but if there are no other options, a loyal New York Giants fan like me will put up with reduced video and audio quality, just to watch the game.

Consumers have become very comfortable with using the Internet for all kinds of daily tasks. Checking the weather, making dinner reservations, and shopping online are easy to do; a few clicks and the information is at your disposal. Two-way communication is taken for granted.

Cell phones and PDAs are beginning to offer many of the features of both television and the Internet. Photo albums, ring tones, and other personalization features build an emotional attachment between a person and their phone.

And what about broadcast television? It has been pronounced dead many times. MSOs were the first assassins, next the Internet, then telcos, YouTube, and on and on. But television is here to stay in one form or another. No other medium can reach so broad an audience with such high-quality production values and immersive presentation.

ONE FOR ALL

The disruption of traditional media consumption patterns is fueled by the transition from analog to digital technology. With the exception of radio frequency modulation, every phase of the content life cycle and every process in media production and distribution is digital.

Fragmentation of media companies based on distribution channels has given way (within the limits of FCC and Congressional regulations) to integrated businesses with a presence on television, radio, Internet, cell phone, and print. AOL's 2001 merger with Time Warner was not ill-conceived; it was simply mistimed.

One factor that was overlooked during the height of the dot-com frenzy was the fact that the explosive growth of the Internet was based on an instantly accessible, pre-existing infrastructure: landline phones. No one (apparently) took into account that in order to deliver rich mixed media, bandwidth greater than 56 Kbps to the consumer was required. Hence, the boom busted when the delivery channel choked.

Internet distribution channels have increased their bandwidth by an order of magnitude or more. However, just as a disk drive will fill to its capacity, sooner or later available bandwidth will be consumed because of consumer desire and demand for new applications.

Fortunately, thanks to the technical innovations developed by the research community, channel capacity has increased while the size of files transferred has decreased. Compression technology, such as JPEG and MPEG, has become household words and enabled rich multimedia, including broadcast-quality audio and video, to be delivered to consumers over channels that could never accommodate uncompressed content.

With the capability to deliver repurposed content, media production has entered a new era: the era of a converged production infrastructure and distribution channel diversity. The conversion that was predicted for consumer media devices has actually occurred at the other end of the media life cycle: production and distribution.

Initially using brute-force, dedicated, linear production workflows for each delivery channel, media organizations are beginning to see the light and are moving to integrated production workflows and infrastructure in support of multiple delivery channels.

WHAT THIS BOOK IS ABOUT

Numerous books have been written about each of the many delivery channels now available to content providers. Each hot topic has its title: *Video over IP*, *IPTV and the Internet*, *Mobile Broadcasting*, *Wi-Fi Broadcasting*, and (my own) *HDTV and the Transition to Digital Broadcasting*, among many others.

Similarly, production has its share of books that describe and detail every aspect of production from the creative, technological, and business perspectives.

Some books have addressed technologies that are new to the digital broadcast operations infrastructure and broadcast engineering. Books that discuss methods used in IT infrastructure design—such as *Service Oriented Architecture*—are available to get media industry professionals up to speed on the latest developments and trends.

This book ties them all together tightly into a real-world framework. Many broadcasters, equipment manufactures, and system integrators already have deployed systems and commission facilities that are creating and distributing content over numerous channels to multiple platforms. As important as it is to understand each production and distribution technology, it is perhaps most important to understand how they all fit together.

WHO SHOULD READ THIS BOOK

When legendary golfer Bobby Jones purchased Fruitland Nurseries in Augusta, GA, in the 1930s, he had an inspired vision for the golf course that was to occupy the acreage. His goal was to create a course that would require a professional golfer to carefully plan and execute a round of golf in order to score well. But recognizing that Augusta National is a golf club, he wanted the members to be able to enjoy a leisurely round. He succeeded, and only the application of advanced technology to golf equipment and the increased training regimen practiced by modern golf pros have forced modifications to the course that better suit modern golf.

I have attempted to apply Mr. Jones' philosophy to this book.

The primary audience for this book is broadcast technologists, engineers, information technology professionals, and software developers. In order to write inspired code and design-efficient media systems, a thorough understanding of how all the technology pieces fit together is imperative. True, a global view is not a substitute for expert knowledge and experience, but a global view helps in system design and fosters cooperation.

However, I've tried to present and explain the material in such a way that non-technologists such as graphic artists, editors, business managers, and executives will find this book readable and informative. Much like a Project Apollo moon rocket, digital media systems are deceptive: simple looking in functional drawings but infinitely detailed in their engineering.

I've also attempted to give technologists a better understanding of the creative and business motives behind production and distribution methodologies and plans. This is an era of unprecedented interdependence of all facets of the media industry, one in which—more than ever before—survival is dependent on looking beyond the confines of functional areas or narrowly defined disciplines to gain an understanding of how all the pieces fit together.

Here, there, and everywhere ... anytime at all

The application of digital technology to the production, distribution, and consumption of media has ushered in the ability to consume content anywhere, anytime, on any device over more delivery channels than any broadcaster or consumer knows what to do with. In order to compete in the diversified twenty-first century media universe, television broadcasters must produce and repurpose content for all possible delivery channels and reception devices.

As if television broadcasters didn't have enough of a challenge converting to digital production and distribution, now they are faced with the reality that this new media universe requires content to be available to on any platform. By 2008, television production companies, national networks, local broadcasters, and cable TV services realized they had no choice but to distribute content over the Web and to cell phones and other mobile reception devices. Yet ways to derive revenue, much less profits, from "new media" have been elusive.

Just take a quick inventory of all the "new media" channels that have become available in the last few years: cell phones, BlackBerries, PlayStations, media center PCs, personal digital assistants (PDAs), digital cinema, digital signage, the World Wide Web, taxi TV, elevator TV, gas station TV, and good ol' print!!! And just to add to the distribution options: Wi-Fi, WiMAX, mobile TV, iPods, iPhones, and don't forget the Internet!

Television distribution also continues to diversify. For more than a decade, over the air, cable, and satellite were the only TV delivery options for consumers. But now, telcos and new media are getting into the act. The acronyms are plentiful—HDTV, RFoG, IPTV, FiOS PC, SDV, VoIP, VoD—even if few consumers know what they all mean.

For broadcast engineers and media technologists adapting to the new technologies and expertise required to design, install, commission, and maintain systems that support producing and distributing content over multiple distribution channels, a new level of complexity has been added to existing broadcast infrastructures. New

1

fields of technology, the integration of IT, and software development have created an infrastructure that is a system of systems. Each area of expertise requires a career's worth of professional competence.

Lost in the need to master esoteric technical details is the big picture, the 50,000-foot view of how all the pieces—the equipment, the production workflow, and the techniques that support the repurposing of content—fit together. This book aims to put together all the pieces of this system-of-systems engineering puzzle.

ONE FOR ALL

Repurposing broadcast TV programming is really nothing new. Retransmission of terrestrial network content over cable and satellite systems has been happening for decades.

Many a television broadcast company owns radio stations as well. And newspaper and magazine publishers have owned TV and radio stations for years. Every media company has consumer-oriented Internet sites; these microsites are often tied to a specific show, magazine, or other niche audience. But things are different now. Rather than having individual business units and separate production departments for each distribution channel, new computer-based systems and a profit-conscious business environment have enabled and necessitated integration of systems and given rise to a mantra of "produce once, distribute everywhere."

Adding to the increasing mass of content is user-generated media. Thanks to the abundance of personal computing power and the proliferation of easy-to-use authoring tools, content consumers have become content producers. Cell phones feature digital cameras and the ability to capture short video clips. Inexpensive production applications format and upload personal productions to social Web sites with a click, and voila, everyone is instantly a publisher with the entire world as the potential audience.

Such is life in the digital media universe.

Digital = Power

Digitized audio and video can be reformatted, converted, and repurposed for any possible delivery channel at a quality level beyond the capability of analog technology. Couple this capability with the rapid deployment and consumer uptake of each new delivery channel, and 2009 marks not only the end of analog over-the-air television in the United States, but also the birth of a new multiplatform global broadcasting paradigm.

Content is now targeted for global distribution. For the Internet, the multiplatform capability already exists. For digital television (DTV) it involves reformatting content, and the 25/50 Hz vs. 29.97/59.94 Hz video frame rate issue has been augmented by compression codec transcoding requirements.

There are two technology variables that enter into the digital universe transformation equation. The first is the transmission of digital audio and video content over a limited data capacity delivery channel. The other is the development of digital production processes and workflows. This book will delve into both and demonstrate that there are many solutions to this system of equations.

Digital television: the ultimate digital challenge

The doorway to this new universe began with the ability to digitize television, or more specifically, video content. Conventional wisdom in the late 1980s said that the ability to digitize a high-resolution TV image so that it could be transmitted over a standard 6, 7, or 8 MHz channel (depending on the national TV system) was probably a decade in the future.

For many decades television was the only "show" that could be enjoyed in the home. Over-the-air distribution, typically referred to as terrestrial television, and the advertiser-supported business model had remained largely unchanged since the first broadcasts. There were no VCRs and few reruns; networks could count on a by-appointment-viewing audience. Television was magic!

Terrestrial television's success was its own undoing, as those in areas with poor reception wanted better TV audio and video quality. By the 1970s, a threat loomed on the horizon, just out of the range of terrestrial transmission.

What began as community antenna systems (CATV) in the 1980s to solve reception problems now began to offer nationally distributed programming. Cable networks (more correctly called "services") distributed content across the country via satellite to local cable companies for distribution. Little by little, cable TV distribution and penetration steadily eroded terrestrial audience numbers.

Direct-to-consumer satellite broadcast services were virtually nonexistent in the early 1980s, with the exception of technology pioneers who installed 6′ C-band dishes in their yards. But by the mid-1990s, new digital technology spurred a resurgence of the delivery channel. By then, terrestrial networks and local broadcasters had faded to a third in the percentage of households for each distribution channel.

But terrestrial broadcasters as early as the mid-1980s realized the potential cable TV had for reducing their audiences. In response the U.S. TV industry and government supported an analog high-definition television (HDTV) proposal for a system based on the work of NHK, a Japanese corporation. One motivation was to slow terrestrial viewer erosion by offering something that would be technologically unappealing to cable operators but highly appealing to audiences. Interested in channel quantity rather than picture quality, cable operators liked the fact that DTV allowed them to offer "500 channels" and maximize use of their installed delivery channel capacity, rather than "waste" bandwidth on HDTV.

Another threat in the United States was the growth of cellular services. Although the use of cell phones had not grown to mass acceptance, Motorola and others lobbied for access to radio spectrum frequencies that were not in use by TV broadcasters.

TV was threatened and didn't take well to the idea of losing its "sacred" spectrum. HDTV became a rallying cry to fend off the attack.

Just as this threat was turned back another appeared, but few could foresee just how big a threat this technology would really be. By 1995 the "consumerization" of the previously academic and commercial Internet had just taken off with the introduction of Netscape and Windows 95. Conventional wisdom proclaimed that real-time audio and—surely you jest—video could never be delivered over the Internet to personal computers (PCs).

Still, there were a few enlightened technologists in the TV engineering biz. So when the Grand Alliance (a consortium of corporations and institutions) developed a digital HDTV prototype, the Federal Communication Commission (FCC) mandated the ability to packetize the bit stream for transmission over a computer network: The emerging Information Highway had to include DTV.

In the end, the FCC adopted this prototype DTV technology, documented by the Advanced Television Systems Committee (ATSC) standard, in 1996. Few envisioned then how the transition to DTV broadcasting would become the technological catalyst that would transform the entire content distribution and consumer electronics industry.

NEW UNIVERSE, NEW AGE, NEW MEDIA

At the turn of the millennium, consumer content delivery channel categories—TV, Internet, and cellular—were independent of each other. Distribution channel capacity and content data volume prohibited cross-platform exchange. There was no way to get broadcast TV over the Internet; the technology simply couldn't do it.

In the new millennium, low-quality audio and video streamed from servers began to proliferate in cyberspace. There was just enough bandwidth to download short video clips to a PC. It was possible to fit 2 Mbps video over a 56 kbps modem. One minute of video would (ideally) take 2,143 seconds—close to 36 minutes!

At the same time, PBS and others began to use the gift of a second channel for HDTV as a means to distribute more than the primary digital TV program. They started by adding a multicast channel to a terrestrial broadcast, most often a weather forecast.

Cell phones were still in their infancy and used for making calls or text messaging. Advanced audio and video and even access to the Internet were still in R&D.

Fast-forward to 2008. The repurposing of TV programming is rapidly expanding to Web sites, cell phones, and soon to mobile DTV. Digital technology also enables interactive entertainment, targeted advertising, and personalization of the media experience on all platforms. In the end, the distinction between devices will become increasingly blurred. Television is not just television anymore.

A numbers game

All distribution platforms are dependent on audience numbers for generating revenue streams. This makes business life difficult in the contemporary, multiplatform

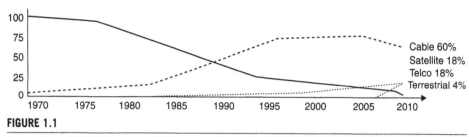

FIGURE 1.1

Percentage of U.S. TV households for terrestrial, cable, satellite, and telco delivery channels since 1970

universe where audiences are fragmented and in control of when and where they consume content.

The migration of terrestrial TV audiences to other delivery channels can serve as a lesson. Unchallenged for over the first 30 years of their existence, terrestrial broadcasters have seen a complete realignment of TV audience distribution over distribution channels that have emerged since about 1980.

Consider present-day TV audience segmentation. Terrestrial television viewers make up less than 15% of the total TV audience. Cable distribution is dominant at over 60% but experiencing churn to direct broadcast satellite (DBS). Telcos are in the TV game now, with fiber optic service (FiOS) and Internet protocol television (IPTV) services, and are gaining subscribers at a slow but steady rate. Figure 1.1 tracks the growth of TV audiences in the United States since 1970 by distribution channel.

In other nations the delivery channel mix varies, but the important point is that TV content is available over a variety of delivery channels. The only similarity among most countries is the progressive disappearance of radio frequency (RF)-based terrestrial television broadcasting.

This is bad enough for TV networks and local broadcasters, but matters have gotten worse. While networks were busy battling with cable and satellite challenges, cyberspace delivery channel capacity grew exponentially. And one day the fledgling, text-based Internet came of content delivery age. By 2007, the rich media Web had become a contender for television audiences.

There are many that will tell you that eventually all TV will come over the Internet. This may not come to pass. But one way or another, consumers will get their content at any time, in any place, on any device.

Technology timing is everything

New media distribution channels, now with adequate and growing capacity to deliver broadcast-quality content along with interactive and personalized features that terrestrial broadcasters could only dream about, have completely transformed the entire media business. As a result, TV broadcasters have embraced multiplatform distribution. As "pipes" (bandwidth of distribution channels, measured in bits per second) to the home get bigger, broadcasting quality content over the Internet is becoming practical if not initially profitable. Delivery of TV content to cell phones, PDAs, and other mobile devices is accelerating.

Around 2005 all necessary technologies coalesced with emerging media consumer trends to reach critical mass. Just as Chairman of RCA and founder of NBC David Sarnoff had to realize that TV is not just radio with pictures, so too did broadcasters have had to realize that the Web, cell phones, and other new media are the future; TV penetration has reached the point of nearly 100% saturation in modern nations and so has little room to grow on its own.

Many factors have contributed to this confluence of delivery channels and the need for integrating platform-specific production processes. Figure 1.2 depicts the relationship among channel delivery technology and consumer adoption. What is important is not so much the actual numbers, but the fact that by 2008, delivery of high-quality TV content is possible to all consumer reception devices.

The resultant multiplatform digital media consumer experience has disrupted the TV and media industry. Internet reruns of yesterday's broadcasts, extra program-related content on microsites, video clips, and cell phone weather updates are available at any time and often for free. Niche services on cable systems enable a viewer to browse for sale listings for homes and cars. You can check movie times and order tickets, and preview an upcoming TV season. The TV experience is becoming more like the Internet.

Media industry business models are changing. The Internet has opened the door to distribution of consumer-generated content to millions with the click of a mouse. Originally nothing more than someone's personal opinion expressed in text and posted to a newsgroup, consumer-generated content on the Internet progressed to

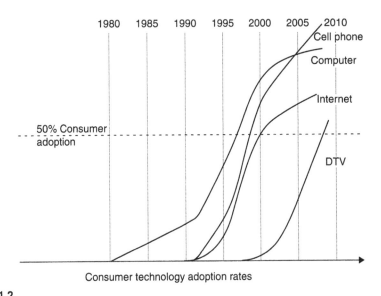

FIGURE 1.2

Technology introduction and consumer uptake of "new media"

personal home pages and sophisticated social networks. The Web has enabled any-one to distribute content on a global scale. Pick the right key words or hit on an interesting theme, and in the course of hours millions will visit your site, or at least your video on YouTube. You're instantly an international star!

The pros get into the act

Up until recently content was produced and specifically tailored for the aesthetics and technical constraints of a particular delivery channel. Radio broadcasts may have spawned TV shows, but radio content stayed on the radio. Print was a world unto itself; maybe a movie or TV screen shot or publicity photo would make it to publica-tion. Analog technology did not support using the same content easily on different platforms. That's all changed.

In a multiplatform universe, content production personnel have a lot more to be concerned about than they did in the channel-specific past. Because content will be repurposed for multiplatforms, talent, writers, graphic artists, and everyone involved in the creative process now must produce multiple versions tailored for delivery over a multiplicity of distribution channels.

But don't get carried away by the romance of new media too quickly. TV still rules the roost and attracts the largest audience numbers. Still, audience preferences are changing. In April 2008, 120 million Internet viewers downloaded a cumulative total of 7 billion videos! These numbers are just too large to ignore: Video quality issues aside, people watch video over the Internet. They find it entertaining.

What began as a wild chase by media companies for new revenue models has settled into a deliberate thrust into new media for both traditional and emer-gent content providers. The wise broadcaster is avoiding being backed into a corner and locked into fading, outdated business models by experimenting with new content formats and new distribution channels. This requires a new take on the media business; this book will show you how the old and new components all fit together.

CONSUMER ELECTRONICS PROLIFERATION

Numerous reports have been dedicated to announcing the death of traditional media such as TV and radio, heralding the Internet as the future of all media distribution. Technology journalism, being what it is, constantly emphasizes the latest and the greatest. But a careful examination of consumer device proliferation and audience consumption patterns paints a different picture.

Mass broadcasting

The oldest means of disseminating content is the "one-size-fits-all" methodology. Content is produced in a single format or fixed medium, and distribution is to the world at large. This was the method followed by TV and radio since their inception. Technology limitations restricted any type of personalization.

I'll say it again: TV is the heavy hitter, period! No other content distribution channel commands audience numbers anywhere near those gathered by an individual TV broadcast. Consider the ratings for annual events like the World Cup, Super Bowl, and Academy Awards. These events are massive global communal experiences and become shared memories, creating a common bond among all who watched.

Audiences are measured by the hundreds of millions of global viewers. In an advertising-supported business, nothing else offers even remotely equivalent reach and commands a similar price per second of commercial time.

DTV, in particular HDTV, has given new life to global consumer electronics. In the United States, the long-awaited analog shutdown has occurred. Global shutdown of analog TV is scheduled over the next five years. HDTV adoption has been steady in the United States and is poised to explode globally. DTV sales have surpassed analog sales.

From November 1998 when the first HDTVs went on sale to 2004, 13.5 million households made the transition to HDTV. By 2010 this number is expected to reach more than 82 million—about 80% of all homes in America.

Program vs. Product

ABC News Chairman Roone Arledge was a visionary. The legendary memo he wrote over a can of beer in Armonk, N.Y. brought "show business to sports." The production and programming concepts and production techniques expressed in those few pages are the foundation of modern television communications. Yet, as he detailed in his posthumously published autobiography, when he heard programming referred to as "product", he knew the business had changed and it was time for him to leave.

TV's founding fathers are nearly all gone. Management has moved on to a second and third generation, people who cannot remember when there was no TV to occupy their time. What was once the viewing audience has become a sea of consumers, to be manipulated and bartered for advertising dollars. Their habits are measured in terms of media consumption.

It is important that content producers not lose sight of their position in the grand scheme of existence as new distribution channels emerge. As Edward R. Murrow said—to educate, inform, and enlighten.

Other forms of mass content do exist. It would be hard to watch a TV program while driving. First radios and then other audio systems have eased the stress of driving for millions. Later, audiocassettes and CDs enabled personalization of the listening experience. Today, drivers "rip" their own fully personalized CDs or load MP3 audio into their car's music server.

There is an ongoing love–hate relationship between the movie industry and television. TV severely impacted movie attendance in the 1950s. In response, the film industry sought technology that would give it an edge over TV. Widescreen theaters and later 5.1 surround sound succeeded to some extent in lifting audience numbers.

Another way erosion of move attendance has been offset is through the sale and rental of video home system (VHS) cassettes and now DVDs. For many, stopping at the rental store became a Friday night ritual. Modern technology is eliminating the need for physical media because movies are available on televisions via video on demand (VoD) and on PCs via download.

Personal entertainment

It's a cliché, but true nonetheless, that change is the only constant; for technology, doubly so. No one can live in a first-world country today without access to a computer and at least a basic ability to use it. Cell phone uptake is beyond 100%.

Personal computing

Since the introduction of the personal computer in the early 1980s, business as well as social and educational endeavors have become dependent on the PC. Almost everyone uses word processing and spreadsheet applications daily at work. At home, the PC pays the bills and stores the family photo album.

The growth of the PC has been exponential. Since the rollout of Windows-based operating systems, increasingly more entertainment applications have been available on PCs.

Initially, PC media capabilities were limited. Computer games, an early "killer-app," featured graphics that were little more than stick figures and backdrops created with a 16-color palette. Clock speeds of 12 MHz and 640 kilobytes of RAM were all a top of the line system could offer. That's nowhere near powerful enough for media applications.

Yet sales of PCs in the United States grew rapidly. It took 15 years for yearly sales to go from virtually zero to 50,000 units by 1995, but in only five more years they jumped to 125,000 and hit 200,000 per year by 2005.

Recent growth has been even faster outside of the United States. Estimates place the number of PCs in the world at a billion or more. It truly is an interconnected world.

An interconnected world

The Internet exploded on the consumer landscape in 1995, and user numbers grew exponentially for the next decade (see Figure 1.3). Even the "dot.com" bust in 2001 did not significantly impact consumer uptake.

The Web is becoming acceptable as a viable outlet for television programming, but video quality remains the pivotal issue. There are innate technical issues that categorically make it nearly impossible for broadcast quality, especially HDTV, to be delivered over the Internet. But viewers are willing to accept low levels of video quality online that they would never tolerate on their televisions.

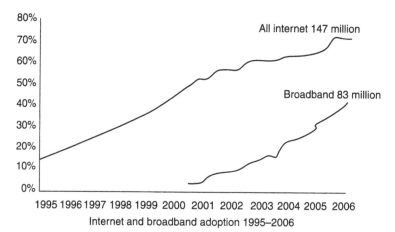

FIGURE 1.3

U.S. Internet and broadband adoption 1995–2006

So how bad is so bad that a viewer will not watch? It's all relative. If the clip I'm watching is just a user-generated video of someone's frog jumping through hoops, it's no great loss. If it's the U.S. Open golf tournament playoff, a dedicated fan will tolerate any level of picture quality, no matter how bad it is.

Yes, content is king, and many would rather watch a less-than-optimum presentation of something they are interested in or find amusing than channel surf the TV only to find nothing they really want to watch.

The Digital Divide and Freedom of Expression and Information

The digital divide is real. Those that stand to benefit most often can't afford to buy the liberating technology. What good is an Internet-delivered discount if you can't afford a PC or access to the Web?

The freedom of expression and access to information in cyberspace that so many enjoy and take for granted is not always available to others around the world. Democracies depend on citizens who are aware of issues and policies, who know what is going on inside the country and around the globe.

Those in power manipulate the media; the media seek the truth but can't help to put a spin on it. Meanwhile, the rich get richer and the poor…well the poor, they just exist and struggle to survive.

One would hope that our new multimedia universe will be used to better the world we live in.

Gaming

The development of PC media capabilities was spurred on by the consumer adoption and business success of PC video games. Competition among application developers was fierce. Improved realism was a marketplace advantage. PC manufacturers saw video games as an opportunity to sell more PCs. Increased effort was expended on developing dedicated graphic processing chips. These integrated circuit cards were incorporated into the system architecture and enabled 3D graphics and near-photorealistic rendering.

With the inclusion of high-end graphics engines, the ability to deliver CD audio and DVD video was attainable. Ensuing generations of PCs included CD and then DVD players. The PC began to supplant consumer electronics.

Free to roam

Reaching consumers with content when they were not at home has always presented problems to media distribution efforts. From a business perspective, employee productivity suffered when employees were away from the office. This was a great incentive to develop mobile phones.

Today, mobility has extended beyond cell phones to mobile computing and DTV.

Cell phones

Untethered verbal communication was a natural extension of landline-based telephony. The meteoric pace of consumer cell phone adoption attests to the need to communicate. Universal cell phone availability has changed the communication habits of the entire world.

As impressive as the U.S. cell phone uptake numbers may be, they are dwarfed by global adoption. This reality adds more incentive to plan to repurpose content anyplace, anywhere, at any time on any device.

From their first consumer introduction, designers of cell phones have striven for higher quality and the addition of more features. 4G (fourth-generation) phones loom on the event horizon, bringing more features and a higher quality video and audio experience.

Portable media-capable devices

Since the dawn of broadcasting people have desired to consume content wherever they are. Even in the pre-TV days, large, vacuum-tube, battery-powered portable radios made the trip to the beach. In time, portable TVs and cassette music players, as well as radio, benefited from the technological advances that enabled miniaturization of electronic components.

Media is all about creating an event. The same holds true for consumer electronics. When the Sony Walkman debuted in Japan in 1979, the media consumption paradigm was forever altered. Previously, portable audio cassette players were about 12″ × 5″ × 4″ high—portable, but bulky. The Walkman could fit in a pocket. Cassette recordings could now be listened to while walking, jogging, or anywhere.

With the limited capabilities of cell phones, it was only matter of time until high-end, handheld mobile computing devices were introduced. These PDAs were mini computers that had limited connectivity to the Internet and could be synchronized with office documents and e-mail.

Given that the advance of technology to support features desired by consumers is inevitable, these early PDAs have grown to the point of including media capabilities. And along the way, cell phones have become "smart phones" and are in essence PDAs. Regardless of what they are called, they can now receive some form of DTV.

Laptops

Use of a laptop computer for media enjoyment has become commonplace. CDs and DVDs can be enjoyed anywhere. Plug in an ATSC DTV tuner to a universal serial bus (USB) port and voila—you can enjoy your favorite terrestrial programs anywhere in reach of a signal. No ghosts or echoes; just perfect digital images and sound. And as wireless networks increase in data delivery capacity, people are increasingly able to view Internet-delivered audio and video from their laptops while away from home.

ACROSS THE MULTIPLATFORM UNIVERSE

For a long while, consumers were perfectly happy to just talk on their cell phones, play games and surf the Web on their PCs, and watch a sitcom on their TV. The idea of doing any task on any consumer device was unthinkable.

Today, daily use of multiplatforms for media consumption and entertainment is spread across a wide variety of platforms. Old-school platforms such as TV, radio, and newspapers have given way to new media channels. Figure 1.4 is a generic breakdown of the number of minutes on each platform spent by a consumer in a single day.

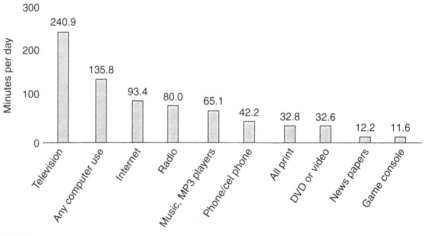

FIGURE 1.4

Media use on an average day by consumers of all ages (Ball State, PEW May 2006)

What is important about this graph is not so much the actual minutes per day spent on a particular platform, but the general trend toward platform diversity.

The impact on content producers, media conglomerates, and business Web sites of the consumers' ability and desire to enjoy media on any platform is transforming content distribution philosophy. A continually increasing number of distribution channels vie for consumer minutes per day.

Personal vs. Professional

Thanks to the *Universal v. Sony* Supreme Court decision, anyone born in the U.S. after 1984 has never been restricted from recording TV content on a VCR. And as the ability to copy audio became ubiquitous, few people gave a thought to the fact that these were copyrighted works, and copies were distributed freely over an Internet-connected world. The distinction between personal and professional creative rights was blurred to the point of imperceptibility.

Content owners have cried bloody murder about illegal copying and consumer copyright infringement. However, they are not so clean as to be justified in throwing stones. In all their concern over lost revenue from illegal copying, downloading, and sharing of media, they have dragged their heels in paying royalties to the writers, artists, and other creative personnel that produce the content they are exploiting.

Audience segmentation

A look at changes in how people get their news exemplifies the general trend in audience segmentation. Newspapers are a mainstay of the information dissemination business, and their status as the ultimate authoritative source went unchallenged for decades. Eventually, evening TV news broadcasts began to alter the news audience consumption platform demographics. Today the Internet has cut deeply into the newspaper industry, disrupting the newspaper business model, as classified ad space is virtually nonexistent. Figure 1.5 traces the trends in channels that consumers turn to for their daily news.

Since the rise of consumer online activity, newspaper readers have diminished from about 58% of the U.S. population to just about 34% in 2008. Internet news access has grown from 0% to 36% and recently surpassed newspapers. Radio has declined to about the same level as the Internet and newspapers. Not surprisingly, the cable news audience has increased to 40% and is in the number 2 position behind local TV.

It's really a no-brainer: News companies must deliver news over the distribution channels consumers prefer. So Web sites and cable outlets are a necessity if a news company wants to stay in the game. Yes, the times they are a (still) changing.

Similar, although not as drastic, audience segmentation is occurring to TV, radio, and magazine consumption audiences. The Internet is mostly to blame, by offering access to niche radio programming, broadcast TV programs, and the entire library of amateur content.

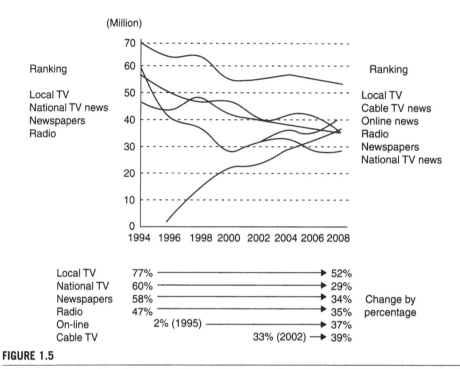

FIGURE 1.5

Change in consumption of news over time on different platforms in the United States (Ball State)

Distribution diversity and the emergence of new media

It was the policy of the Clinton administration to deregulate the telecommunications industry. And as the industry was deregulated, and long-standing legal impediments were eliminated, media megaconglomerates came to be.

As the Internet rose in household penetration and in the delivery of multimedia content, the merger between AOL and Time Warner represented the first step toward channel integration. The prospect of a unified broadband future was just around the corner, according to many media technology and business analysts. Yet by any metric, the merger was a failure.

One could argue that the consumer success of the Internet was possible because the interconnection infrastructure was already in place. Just buy a PC and plug it into your phone jack and you were cyberspace-enabled. What few considered was the time and expense necessary to roll out a new broadband infrastructure to support a multimedia Internet. So bad timing may have been the underlying cause for the failure to realize the vision.

But time has proven that the vision of a multiplatform, multichannel universe was correct. It just has taken longer than expected to get there.

Branding

Marketing revolves around brand awareness. Consumer loyalty and relationship-building require a constant presence of brand. This reality was so important to terrestrial broadcasters in the United States that after the DTV standard was adopted by the FCC, the ATSC developed an additional standard to ensure that after the analog shutdown and the relocation of channels to new frequency slots, viewers could still identify and navigate to programming based on the old analog channel number.

As broadcasters move into new delivery channels, customer brand loyalty can directly influence the success of any new channel. Advertisers can be offered a mix of presentation platforms with refined demographic characteristics. So, regardless of the fact that these new distribution channels have yet to turn a profit, they are absolutely essential to brand awareness.

TOWARD INTEGRATED PRODUCTION

The digitization of audio and video created an opportunity to easily exchange content between consumer devices. Pristine copying has precipitated rights management issues, and preventing (or at least discouraging) unauthorized use is a defining factor in any content distribution strategy. No segment of the media industry wants to repeat the Napster fiasco that nearly destroyed then rewrote the rules of the music business.

Islands of production for each delivery channel have been the status quo for decades. Spinning off of a promo or tease from program material has been but a small part of broadcast operations.

To remain in that mindset is certain death for any content-producing organization. As this book will explore, today content producers, broadcasters, and distribution channel operators must think in terms of the big picture and leverage their production and distribution infrastructure to maximize usage and emerging new media revenue opportunities.

Linear platform-specific production

With the increasing need for compelling content to attract audiences, production and creative personnel are being taxed. The plethora of consumer platforms and commercial signage that must be supported has led to increasing demands on existing production personnel, workflows, and equipment. The answer is for organizations to do more with the same number of people. In other words, they must leverage digital technology and workflows for improved efficiency.

Faced with the large capital expenditures necessary to convert to digital production and broadcasting, it is easy to understand why there is a need for the design and implementation of integrated production systems. Linear, single-channel workflows

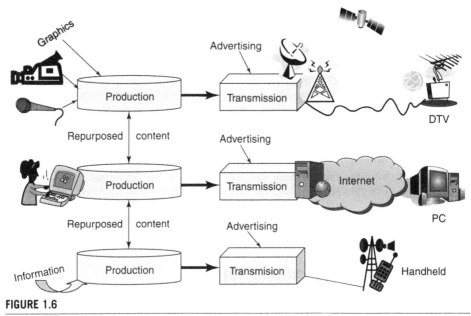

FIGURE 1.6

Independent production and transmission workflows for each consumer device

and infrastructures will get the job done, but at a cost in time, effort, and expense that is not acceptable in today's business climate.

With the rush to get content distributed over new media channels, brute-force production assembly lines offered a quick solution. As shown in Figure 1.6, a discrete production process fed each delivery channel. Content was created and formatted independently for each reception platform. When the need arose to repurpose existing content for another delivery channel, existing clips and/or data were fed into the assembly line for the target delivery channel.

New outlets such as digital signage in stores and custom broadcasts to elevators, gas stations, and taxis are increasing the strain on production resources. This content must also be converted to transmission formats that are compatible with diverse distribution channels and multiple consumption devices. Production requirements will only intensify as new platforms and delivery channels continue to come into existence.

Converged parallel production

Innovation is a given in the media business. New technologies that enable creative, business, and infrastructure enhancements are constantly sought. As the world went digital, so too did content production.

However, in many instances, existing linear production and distribution workflows were maintained during the transition to digital production and multiplatform

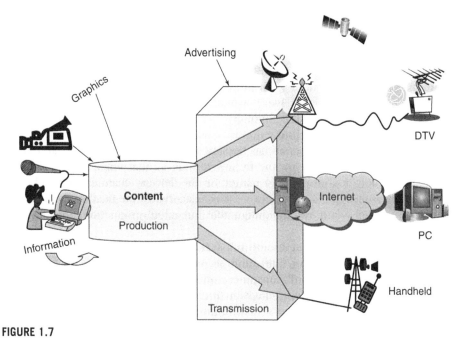

FIGURE 1.7

An integrated multiplatform production workflow

distribution. But as more distribution channels and more consumer reception devices appeared, the ability to reuse content on all channels became a driving design requirement. Those with foresight realized it made good business and technology sense to coordinate the linear processes. Why repeat common processes? The concept of integrated production was born.

The key to integrated production is the identification of technology and workflow overlap. Figure 1.7 presents a simplified view. There are quite a few production processes that are common to all delivery channels. Intelligent design of a production infrastructure will implement a system that can produce appropriately formatted content for the transmission stage of the broadcast chain for each delivery channel.

Workflows for content production and presentation share numerous common processes. With careful analysis and thought, streamlined production that overlays the technical infrastructure can be developed.

A nonlinear world

Abandoning video and audiotape, TV production has moved to file-based production. This enables the simultaneous production of content in all the formats required for distribution over multiple channels. Cumbersome linear tape-based editing takes a lot more time than nonlinear editing (NLE). With a NLE workflow, more than one editor can access raw content at the same time. Keep in mind, digital copies are perfect, so production values remain pristine no matter how many generations of editing are performed.

Digital broadcasting has led to the influx of massive amounts of IT equipment in production infrastructures. PCs and media servers are everywhere in a Broadcast Operations Center. A PC running an automation program often controls programming and commercial insertion. Every aspect of production in all media businesses has been transformed by digital technology.

Because production must now support a wide variety of delivery channels for multiple devices, format conversion has become a huge issue. Transcoding and graphics compositing must be done on the fly, in real time, so that when content leaves a master control room, it is properly formatted for the delivery channel and can be presented on the associated reception device. This exacerbates the need to use common processes in parallel workflows in configurable, integrated production control rooms, and master control rooms.

Individual pieces of equipment and full system solutions that address integrated production, diverse distribution, and multiple receiver requirements are beginning to appear. Both professional and consumer content creation applications can generate audio and video formatted for television, Internet, and cellular channels.

Integrated production for diverse distribution channels to multiple devices isn't about working harder; it's about implementing technology and workflows that work smarter.

AN AUDIENCE OF CONSUMERS

There is seemingly no end to where digital media will be available for consumption—elevators, taxis, gas station pumps, and malls are but a few of the heretofore inaccessible places that media now reaches. It is safe to say that there is virtually no limit to where media will be consumed, save the requisite technology to get it there.

And viewers are not only getting content where they want it, but also when they want it. Today, time shifting is the norm for many viewers. Many television set-top boxes (STBs) have incorporated digital video recorders (DVRs). If a station offers five show times instead of one, will it add to or erode viewership? How will this impact advertising? If the audience is reduced by 10% for the first run, will it be offset by DVR viewing? Will commercial skipping cause advertisers to lose 10% of their audience? Can this be offset by Web rebroadcasts? Can technology disable commercial skipping? Is there a way to enable commercials-on-demand?

Advertising is about increasing brand awareness and sales. Any means to this end is the goal regardless of the media mix. This dilution of viewers could be offset by developing a personalized, targeted, multiple consumption device strategy. Viewers are directed from platform to platform for more interesting content. In a sense, this is a form of viewer relationship management.

These are business issues. Although profitable models have not yet been discovered, one thing is clear: All broadcasters must deliver content over all possible distribution channels and must try new approaches to revenue generation. As always, the ones who are most successful at adapting will survive.

However, there is a riddle to solve: What is the equation that will produce a favorable value proposition for all involved? Like Oedipus and the Sphinx, if a media company fails to solve it, its competitors will devour it.

Just as in 2001: A Space Odyssey, where the planets aligned, the monolith was discovered, and humankind took its next evolutionary step, so too has the media business made a quantum leap into a brave new world. Digital technology is the monolith.

"The meek may inherit the earth, but only the bold will ever find heaven"

from "Opportunity Knocks"
by Phil Cianci copyright 1995 Iopherian Creations

2

The multiplatform experience

For most of the history of broadcasting, the audience has had one choice: either watch or listen to a broadcast when it is scheduled or miss it. TV viewers sometimes got lucky, as summer proved a perfect time to rerun a select number of previously aired programs. Radio listeners rarely, if ever, had an opportunity to hear a rebroadcast of any kind.

So-called appointment viewing was disrupted by the invention of the videocassette recorder. In the 1984 landmark *Universal v. Sony* legal action, the issue of personal recording circumventing copyright rose to the Supreme Court. By the slimmest of margins (5–4), the court ruled that the concept of "time shifting"—making of a copy of a program, for personal use, to be viewed at a time other than its original broadcast time slot—was fair use.

A side effect of that decision was the eruption of one of many consumer marketplace technology battles to come—the VHS/Betamax war. Ultimately that fight was decided by convenience rather that technical supremacy. Both were analog technology, so each generation (copy of a copy) suffered a degradation of audio and video quality due to the successive addition of inherent electrical noise with each copy. Betamax offered superior video quality and continued to be a mainstay of professional video, but the two-hour capacity of VHS served both consumers and Hollywood studios better. The precedent had been set; consumers were now taking control of their media consumption, and this wouldn't be the last time quality lost a battle in the consumer marketplace.

As new consumer technologies rolled out, in particular the Internet and cell phones, "consumption" of long-form programming (a half an hour or more) remained limited to television and radio. Neither of these new technologies had sufficient capability to receive and present audio and video. Both distribution channels had severely insufficient data transfer rates.

But this was soon to change.

TECHNOLOGY CONVERGENCE

In the early 1990s, before the Internet was a household presence, the National Information Infrastructure (NII) initiative, promoted by the U.S. government, had a strong influence on the development of digital television. A winning system in the HDTV standardization war was expected to deliver more than video and audio, and be compatible with what would soon be dubbed the "information highway." Data services were a system requirement, and as per FCC Commissioner Alfred Sikes' far-sighted March 1990 decree, digital HDTV would be compatible with this emerging network. One competitor in the HDTV standardization war announced later that year that it had a system that could accomplish the feat.

Audio and video had been digitized by this time, CDs were everywhere, and many processes in TV production were now done in the digital domain. The amount of data necessary for digital delivery of real-time digital audio and video was at least an order of magnitude beyond what any existing distribution channel technology could handle.

Work was underway to "compress" audio and video, using a combination of existing data reduction techniques and new methodologies. In the end, the Moving Picture Experts Group (MPEG), a professional working group comprised of video and audio researchers and engineers, figured out how to process raw audio and video to fit on limited storage media and played back in real time. MPEG-1 hit its target of VHS quality, and the DVD came into being.

But there was still a lot more work to be done for digital television to come into being. The MPEG-1 "tool kit" needed to be expanded to include capabilities for real-time, broadcast-quality audio and video. Digitization techniques had serialized analog TV signals in the mid 1980s at data rates in excess of 100 megabits per second (mbps), way beyond the capabilities of any extant digital delivery consumer channel.

Clever researchers accomplished the quest. MPEG-2 was capable of broadcast-quality standard-definition (SD) audio and video. This still left HDTV out in the cold. MPEG-3 was meant to crack this riddle. By including a few more data reduction techniques, broadcast-quality high-definition video and 5.1 channels of CD-quality audio could be compressed to under 20 mbps, an astounding 50:1 data reduction without apparent image or audio degradation.

With the problem of how to reduce hundreds of millions of bits a second of video and audio into a few megabits occurring at the same time as the NII initiative, the convergence of the television and PC was predicted to be imminent. Microsoft, PC manufacturers, and the MIT Media Lab pushed the convergence concept to the point that the union was a done deal.

This led to the idea of an interconnected home media environment and was a hot topic of industry discussion in 1995. At the EIA/IEEE (Electronics Industry Association/ Institute of Electrical and Electronic Engineers) convention in Philadelphia that year, attendees discussed a Network Interface Device (NID) that acted as a media gateway into a home media network. The Internet was an add-on.

PC manufacturers saw things differently; naturally, the PC was at the center of an Internet-enabled media universe. Yet even Microsoft's Bill Gates had to admit in

1997 that the convergence of all media consumption with the PC at the center was not quite the right scenario. Coalitions including the PC camp had successfully lobbied for the removal of 18 video presentation formats from the DTV standard to ease their entry into the TV business. Gates had to admit that the vision discussed in Philadelphia—that there would be three modes of consumption, passive (television), active (PC), and personal (handheld)—was correct and that the PC would not be the center of the media universe.

At about this time Netscape was introduced, followed by Windows Explorer's inclusion in the Windows 95 operating system. The Web browser wars and legal action began. Consumer use of the Internet grew rapidly, mostly because of the fact that every home had a phone-line connection; dial-up modems with up to 56 kilobits per second (kbps) were the rage.

As content became rich in audio and graphics, it became painfully obvious that 56 kbps data transfer rates were excruciatingly slow. Cable multiple system operators (MSOs) got into the act and began to offer increased Internet speeds, leading to today's broadband data rates of multiple megabits per second. Telcos followed with high-speed digital subscriber lines (DSL), and the battle for customers still goes on.

Today, real-time video streaming of compressed TV content over the public Internet at close to SD broadcast quality is a reality. User-generated content sites such as YouTube and MySpace have "democratized" media publication.

Easy access to sophisticated creative tools has improved the quality of home productions. But the distinction between personal and professional is often lost; it's one thing to record a TV show for later viewing, and something else to use it on your social networking page. This paradigm shift threatens to rewrite copyright law in the name of information exchange and technological innovation, a topic the second part of this book will discuss in detail.

A MEDIA-CENTRIC EXISTENCE

Electronic communication technology has long mystified the public. Being able to send a message across great distances was seen as a miracle, or magical at the very least. People jumped when a train rushed at them on a movie screen. Women fainted as Harold Lloyd hung from a clock on top of a skyscraper. Contemporary minds could not process the images as anything other than real. Yes, it was movie magic, the illusion of reality.

Theatrical presentation of movies on the "big screen" created a larger-than-life experience. When television emerged, the best that CRT technology could offer was a small black-and-white display of less than 12 inches. Cinema set the expectations for visual communications, and the newly invented television paled in comparison.

However, a funny thing happened, something that seems to happen over and over with the introduction of new media presentation and consumption devices. In spite of the poor quality of television as compared to the cinema, movie attendance seriously declined as post-WWII America purchased increasing numbers of TV sets.

Ease of use, the comfort of home, and the fact that advertising subsidized the content, rather than consumers paying for it directly, won out over quality.

Media Psychology

Why do people watch television or listen to music, or read a magazine, and what is the impact? Stuart Fischoff, Ph.D., offers his thoughts in an article titled, *Media Psychology: A Personal Essay in Definition and Purview.*

He begins by defining the discipline:

"Broadly speaking, media psychology uses the theories, concepts and methods of psychology to study the impact of the mass media on individuals, groups, and culture ... with the social and psychological parameters of communications between people ... that are mediated by some technology or conduit other than simply air."

On the Pervasive Effects of Media

"The media shape the way news gathering and transmission, advertising, political processes, and campaigns, wars, diplomacy, education, entertainment, and socialization are conducted. The effects the media have on these human enterprises are legitimate points of interest on the expanding scope of media psychology."

On Media and the Sense of Self and Self-History

"Existential ramifications are dizzying in an age of media innovation, expansion, and penetration. As media content and technology evolve, so does culture. As culture evolves, so does the media and so does the manner in which reality is understood. As culture evolves, social and political agendas are set, and personal exploration, discovery, and productivity are conceived of, formed, and transformed, in an endless developmental cycle of technological breakthrough, social adoption, market penetration, maturation or saturation, market decline, and, eventually, adaptation ... in the face of newer technology and attendant competition for consumer dollars and consumer attention."

He concludes:

"Arguably, the field of media psychology may be the study of a religion. The forces of media create the celebrity gods we both adore and hate. They create the means by which we come to understand ourselves and evaluate others. They provide the intellectual, spiritual, and hedonic manna that fills our senses, and alternately crystallize or cloud our thoughts. They inspire the dreams of our ambitions and the demons of our nightmares."

[Excerpted from *Journal of Media Psychology*, Volume 10, No. 1, Winter, 2005: http://www.calstatela.edu/faculty/sfischo/MEDIADEF-2.html (online publication date February 8, 2005)].

Three screens

Media consumption is a habitual activity for many people. Turning on the television the minute they wake up, switching on the radio as soon as they turn the ignition key in a car, or reading the morning paper over breakfast are as ingrained as breathing.

Observation of media consumption habits has led to the classification of functionality based on screen size, device functionality, and consumption environment. The classifications are frequently referred to as a three-screen scenario (Figure 2.1).

Television is a passive, lean-back viewing experience. Big screens and surround sound bring a theatrical experience to the home. The audience is at least three-screen heights back from the display; for a 40-inch diagonal HDTV, picture height is about 20 inches and optimal viewing distance is 60 inches back from the screen.

Active, lean-forward consumption is a characteristic of computer use, an outgrowth of the earliest word processor, the typewriter. The user is at a keyboard and mouse less than 18 inches from the screen. In the beginning, audio and video were secondary attributes; it's a text- and image-based information experience. But that has changed, and today's home computers are powerful enough to play audio and video.

Personal communication devices such as a cell phone are the third mode of communication. In 1995, landlines were still the dominant telecommunication technology, and consumers were used to and dependent on the separate, non-110V AC electricity, lifeline capability of the phone system. If the power went down, your phone still worked and you could call for help. Today cell phones fulfill that fail-safe communication need and also feature cameras as well as the ability to play video and audio, and connect to the Internet.

Television

The television is considered a household appliance. Turn it on and it works, flawlessly. High-end television can fill a room with equipment. As Figure 2.2 shows, an HDTV with surround sound takes up a lot of space. Welcome to the age of the home theater.

Screen sizes continue to get larger. New homes are wired for movie theater-like viewing experiences, dedicating an entire room to entertainment. The family room has come a long way in the digital age.

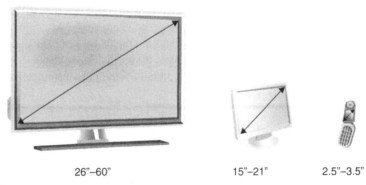

26"–60"   15"–21"   2.5"–3.5"

FIGURE 2.1

The three-screen universe: relative sizes of TV, PC, and handheld displays

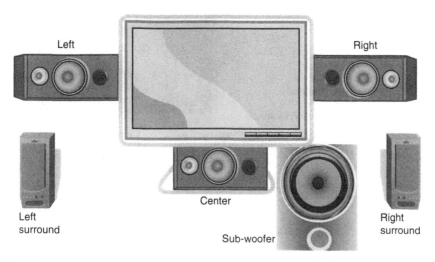

Left

Right

Center

Left
surround

Right
surround

Sub-woofer

FIGURE 2.2

An HDTV home theater system with 5.1 surround sound

TV remote control and onscreen displays

Yet even with all the audio and video quality improvements, a remote control continues to be the way a user commands and controls a television. For the most part, it is one-way communication: the user tells the television what to do.

One complaint from the average consumer about DTV is how difficult it is to control. Channel changes seem to take forever. STBs and displays frequently get out of sync; one turns on, while the other stays off. Even with the effort to simplify the user interface and make control functions user friendly, many viewers avoid learning how to take advantage of even the most fundamental remotely controlled features.

Control functionality for DTV receivers fall into two categories: setup and content navigation. Most viewers who are inclined to do so have gotten used to adjusting the color or brightness using the remote and an on-screen display. The difficulty with DTV is all the additional choices. Aspect ratio and signal source options can drive a viewer mad. Many viewers just accept distorted images, too many buttons to push, hitting the wrong button and accidentally changing the signal source, a blank screen, and reading the manual, a very frustrating experience. So much for being an appliance. And that remote control—what do you do with 50 or more keys?

Development and standardization of a mechanism to find and to tune to a desired program may have killed the TV-listing business, but they are a bright spot in the transition to DTV. Information contained in a transmission provides technical and descriptive data about the program. By assembling all the available information about upcoming programs, an on-screen guide can be constructed. Figure 2.3 shows the result: an electronic program guide (EPG).

EPGs offer myriad capabilities. The program grid displays programming based on attributes such as channel (shown in Figure 2.3). By using the navigation keys and

FIGURE 2.3

Generic electronic program guide (EPG) user interface

moving the cursor highlight, the viewer can get more info, use record features, and change the channel.

But the real power of an EPG—the development of features that support locating programs that you don't know exist—is still in the future. Right now this is rudimentary. EPG browse options are limited to channel, theme, or name. Some systems can perform limited program searches based on user input, but the implementation is a far cry from the search capability we have all become accustomed to on the Internet. The limited features offered by current EPGs are grossly inadequate and fail to meet user desires.

Up until now, an STB was necessary to provide EPG features. EPG implementation for terrestrial broadcasts had been nonexistent. DTV manufacturers have recently begun to include program-listing features in their standalone receivers.

Computers

Computers are rapidly becoming the center of the home media environment. Many of the features present in a DTV receiver are present in a computer system. If a DTV tuner is included in the PC, I can compute and be entertained all at the same time. If I see something I want to know about in a show, I just minimize the TV window, open a Web browser, and search away over the Net.

Home system components

Although the packaging still looks much as it did 20 years ago, home computers are extremely powerful and sophisticated. A generic system is shown in Figure 2.4.

FIGURE 2.4

PC System with 2.1 sound

It doesn't look like much, but the display is most often HD capable. The connected sound system rivals dedicated audio systems.

However, home computers integrate many, if not all, of the features in a home theater without the need for multiple devices, hence the appellation "media PC." And if one is adventuresome, an HDTV and surround sound system can be connected to the PC, combining the best of both experiences. Some will argue that a computer delivers a significantly better media experience than a television, with video, music, photos, games, and information, all accessible from a single user interface and device.

GUIs and windows

Computer graphical user interfaces (GUIs) are well developed, and most people feel quite at ease using them. Available functions are logically grouped on the screen. With the elimination of the command line interface (many years ago) as the primary means of control, computer operation has become intuitive. Drag and drop, left click and right click actions simplify every operation. How did we ever get along without a mouse?

Figure 2.5 presents representative features of a windows-type user interface (UI).

Multiple tasks can be run simultaneously in different windows. Applications are invoked with just a click.

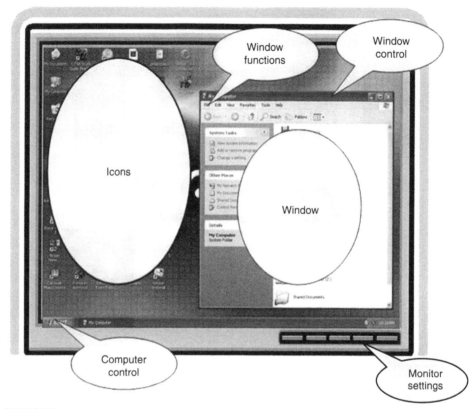

FIGURE 2.5

The organization of a typical computer user interface (UI)

When a home computer is equipped with a media center application, in many ways its functionality surpasses a DTV or STB. A media center PC top-level menu screen is shown in Figure 2.6. Every choice of media experience is available simply by selecting and clicking.

A media center PC program guide looks nearly the same as EPGs presented by an STB. The similarity ends there. If I so choose, I can size the windows on the display so that I can compute and watch television at the same time on one screen. I can surf the Web. And maybe most importantly, when a TV program or advertisement tells me that more information is available online, I can go there immediately.

Mobile and handheld devices

But you can't use a television or PC everywhere. This is where the third screen excels. I can keep it in my pocket, hold it in my hand, and watch it anywhere at any time. I can even talk to other people with it. The size of a handheld device is no more than a few inches wide and a few inches long, and is less than an inch thick.

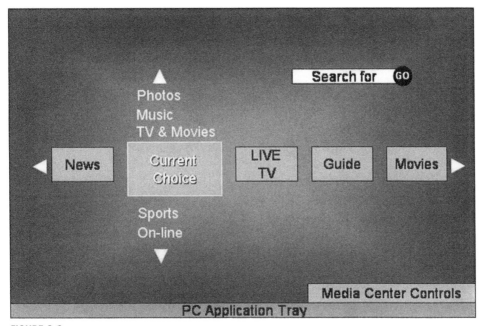

FIGURE 2.6

Typical personal computer media center user interface

Cell phones and PDAs

What the Internet was to the dawn of the information age, cell phones, personal digital assistants, and other handheld untethered devices have been to the spread of geographically ubiquitous communication and entertainment. The various form factors and user interfaces of contemporary handheld devices are shown in Figure 2.7. They can be divided into categories: cell phones with a traditional pushbutton array; those with keyboards; those with touch screens; and those with both, full QWERTY keyboards *and* touch screens.

Cell phones and personal digital assistants continue to add new capabilities. The latest generation of devices offers Internet access and TV programming.

Handheld user interfaces

Cell phone feature navigation is very similar to computer navigation. A select, click, and drill-down navigation paradigm is used. Figure 2.8 shows successive screens in a typical use scenario.

Suppose you have the latest and greatest smart phone and you're hungry. You click the icon for a restaurant locator. Using its global positioning system (GPS) to figure out where on earth you are, the phone displays a list of local restaurants. You vet the list based on the kind of food you'd like tonight. A reduced list of choices appears, you select the one you want, review a menu, and press select to make a reservation. The phone places the call, and in a few seconds, you're booked.

FIGURE 2.7

Cell phone forma factors and user interfaces

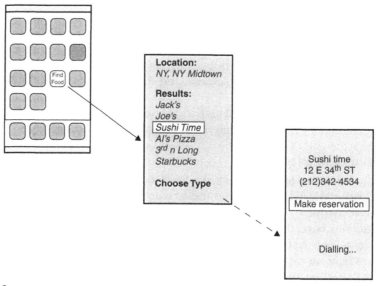

FIGURE 2.8

Typical cell phone menu navigation

A homogenous experience

Televisions, home computers, and handheld devices are all becoming more and more alike. Every feature and every type of entertainment is available on each of them. With little difference in capabilities, the selection of a consumption device defaults to user preference or convenience. This is media consumption freedom. An individual is not forced to choose one device or another; rather, the consumer can have several different devices, choosing to view or listen to content on one in a given time and place, while selecting a different device for another.

PRESENTATION CHARACTERISTICS OF EACH SCREEN

No matter how much promotional propaganda is disseminated, there are too many variables involved and new technologies waiting to be implemented, and the marketplace is too fickle, for anyone to even make an educated guess about what the future holds for media distribution and consumption. But I'd be willing to venture this prediction: in one form or another, all three screens are here to stay.

Obviously, the media experience is different for each of the three screens. Different types and amounts of information can be conveyed to each type of device. The presentation characteristics of each platform influence the way content is produced by a broadcaster. Channel data rate has the final say; if it can't fit in the pipe, it can't be delivered.

The "big" screen

Television has evolved from just video and audio presentation to the point of distraction. So much information is often crammed on a screen that it is often nearly impossible to enjoy the programming. In other circumstances, the added information has real value and enhances the media experience. It is a wise producer who knows the difference.

Figure 2.9 is a generic screen that shows many of the visual elements that are used in a contemporary TV broadcast. A crawl is the moving text usually at the bottom of the screen. Most of the time, it presents information that is not directly related to the current program. A logo, either network and/or local ID, is frequently displayed in a corner. In the United States, FCC-mandated display of the parental rating is presented at the beginning of a program and after commercial breaks. Closed captions are only displayed upon user request, but the information must be contained in the broadcast.

During a broadcast, snipes will appear in the corner promoting an upcoming show. Generally, they are animated, video vignettes designed to grab your attention. Newscasts feature the over-the-shoulder shot; sportscasts display clock, score, and other game information. Billboards are static presentations with a voiceover. A viewer can get lost in too much information (TMI).

FIGURE 2.9

An information-laden TV display; lower third, crawl, bug, snipe, logo, and pillar bar data

Regardless of the delivery method—over the air, cable satellite, or telco fiber optic to the home—the presentation components are the same for each delivery channel. This is what television is.

The "active" screen

A computer offers a different set of characteristics. The display, even if larger than 22 inches, is considerably smaller than most televisions. The audience, generally referred to as the "user," sits within 2 feet of the display. By its very nature, the computer is an active device; it is told (commanded) by the user what to do via the keyboard, mouse, touch screen, or even by voice command.

A DTV tuner can be installed on a PC to receive television and radio over the air, or via cable or STB; video can also be delivered over the Internet, and a DVD will play back just as well on a PC as it will from a dedicated DVD player through a television. In any case, the computer can display the video in a window or in full-screen mode. Multicore, gigabit clock speeds have brought this within the reach of most new computers on the market today. The viewing experience will be nearly identical in quality and content, including graphics and text, as it would be on the TV screen. Sound is dependent on the audio card and installed speakers, but most PCs support surround sound as well as conventional stereo.

Full-screen viewing certainly has its appeal, but the PC media experience comes into its own when the video is displayed within a window among other information and interactive options. Figure 2.10 shows the elements of a typical Web site that includes video.

When the site is opened, video appears and plays in a window, usually with media player controls. While the video is playing, the user has access to all the usual Web site features. The content provider can place thumbnails off to the side that enable the user to select from a list of video clips. Site search and Web search are available. It's the Web and audio and video programming all at once—a technophile's heaven.

However, this is not a TV experience, and there are presentation issues. In a windowed video, a TV image on the PC display may have difficulty presenting legible graphics and text information. On the other hand, a full-screen image may come close, but will not attain broadcast quality.

On the plus side, additional information and advertising can be displayed simultaneously when not in full-screen mode. Web-based targeting technologies, involving cookies, history, and profiles, can be used to display relevant ads, just like when a user is visiting any other Web site.

Advanced CD and DVD features are available on an Internet-connected computer. A short while ago, if you placed a CD or DVD in your computer drive, about all you would get was a playlist. Today, all kinds of additional info pops up. You can even buy more content by the artist or media company. All this is enabled by the fact that the Internet has developed secure, e-commerce capability.

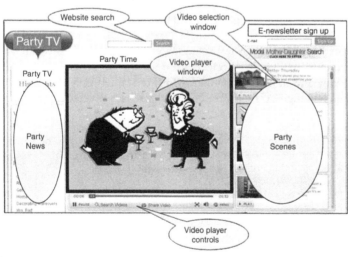

FIGURE 2.10

Web browser interface with video player with interactive "extras"

The "mini" screen

A cell phone offers many benefits but creates another set of challenges for content delivery and presentation. On the plus side, you can privately consume content anywhere (provided you have a pair of earphones). News and sports updates are available. 3G smart phones are more like mini computers with Web browsers than cell phones.

But video is severely downsized. Forget about seeing graphics and text that has been assembled for television. As Figure 2.11 shows, there isn't a lot of room to display information. Display processing circuits that convert a video frame for presentation on the mini screen will filter out much of the detailed visual information present; this generally includes smaller font text and some graphics. To get around this, the graphics and text must be processed separately from the video. This presents production and transmission challenges, as will be discussed later, for any TV broadcaster who wants to repurpose programming and distribute it to a cell phone.

Audio presents a different kind of a problem. Sound is reproduced by a mini speaker or earphones. The speaker is only powerful enough to be heard when it is held close to the ear, and it sounds tinny; the earphones keep the experience contained and the frequency characteristics improve, and no one nearby has to listen to your soundtrack.

Media consumption on a cell phone usually is by necessity rather than choice. Would anyone choose to watch the Super Bowl or a first-run drama on a cell phone if an HDTV were available? I can't imagine why. But if I'm on a train or somewhere else away from a television or PC and my team is playing for the league championship, I don't care about image or sound quality, only that I can perceive enough sensory stimulus to follow the game in real time.

Now consider a "triplecast" to all three types of devices originating from a broadcast operations center. In order to use audio, video, and graphics on all the platforms, each of those elements must be reformatted for the presentation capabilities of the reception device and fit in the data bandwidth of the delivery channel.

This is the challenge of the three-screen broadcast universe.

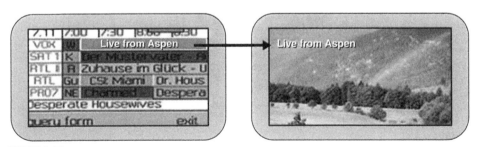

FIGURE 2.11

Cell phone EPG presentation grid and program selection

A DEBT OF GRATITUDE

Although this book is television-centric, due homage and respect must be paid to two sectors of the media industry that will be mentioned from time to time. Without the invention of these enabling technologies, the modern media universe would not exist.

The Wizard of Menlo Park may have illuminated the world, but he didn't stop there. He cleverly devised methods to record sound and to create the illusion of motion from still images. Thomas Edison can rightfully be called the technological father of both the audio recording and motion picture industries.

Sound without pictures

Sensual perception and the impact of stimulus varies for each of the five senses. Some people have photographic memories; others can remember a smell for decades. The feeling of the warm sand on a Caribbean beach may never fade, and the taste of Mom's chicken soup is something that will always be with you.

Hearing and music are of the ears. There is a certain type of spiritual aesthetic to music. Similarly, the spoken word has profound impact. Topics such as politics and religion evoke primal responses. There is truly something unique about reception of vibrations in the air by the ears and the processing by the mind that follows.

In the media business, sound recording has been and continues to be a world unto itself. MTV may have added a visual element, and pop concerts are not complete without video screens, but people listen to music for the sake of the sound and its effect on the soul.

It wasn't long before technology improved on the original cylinders, and flat platter recording technology and mass pressing upped the ante. Musical recording became big business. In time record players were in many a household, playing the latest and greatest hits. Music had become a staple of the masses. One could listen to the great Caruso while sitting in a favorite easy chair.

In the days following WWI, radio transmission was an established means of commercial communication. Every ship at sea was equipped with a radio and regularly communicated with shore stations. Eventually, someone got the idea of playing a record over the air; the world has never been the same since.

Records held their musical throne for many a decade through their evolution from 78 rpm to 45 rpm to 33 rpm, and longer and longer playing times. Even to this day, audiophiles will swear by the sensual aural superiority of vinyl.

Audio cassettes were convenient and mobile. However, their fidelity shortcomings were obvious to all.

Digital technology outperforms analog recording and pressing period. All romantic aesthetics aside, the dynamic range and timbre accuracy could only be approached by professional recording studios using analog technology. The real problem today is that sound equipment is not engineered for fidelity. Market research determines what consumers like best, and products are designed and manufactured to meet customer preferences.

Compact disc technology revolutionized the music distribution business. CDs, based on optical disc recording, were superior in nearly every way to records. When they appeared, it seemed like they would have a long, prosperous reign.

By the 1990s, digital audio production equipment had dropped sufficiently in price to be within the reach of all professional recording studios. CDs that featured end-to-end digital production put an end to all but a few die-hard professional analog audio businesses and to consumer analog distribution formats.

As bright as the CD nova was, it was relatively short lived compared to its vinyl predecessor.

The near perfection attained by digital audio recording and CD distribution gave way to convenience. The combination of home computer proliferation and improved audio compression led to the next major audio development: distributing music over cyberspace.

CDs did not include any form of copy protection. Once CD drives with write capabilities were on every new computer, it wasn't long before people were copying CDs en masse. This ability to burn CDs would prove to be the assassin of the CD itself.

The Internet, with its need for minimal data rates and maximum quality, was the catalyst. As audio compression codecs found their way into consumer PC-based recording applications, the next logical step, after ignoring copyright, was to compress CD audio and share it with a friend over the Internet.

When the dust cleared, the music business was completely transformed.

Images with motion

From its initial invention at the turn of the twentieth century to the beginning of WWII, motion pictures were the only show in town. During this time, many new technologies were implemented and the movies progressed to include sound, and then color. Try to think of the reaction of an audience, accustomed to black-and-white films, when Dorothy opened the door to Oz awash in vivid color on the big screen.

And then came television.

The film industry has a long history of being at odds with the TV industry. The repurposing of Hollywood films for network television, however, aided the movie industry's recovery from the impact of audience migration to their living rooms in the 1950s. There's no holding back progress, and today movies are distributed on subscription-based cable and satellite systems, DVD rentals and purchases, and via the consumer-controlled world of VoD. Movies are available on all three screens.

Many cinematographers still prefer the "film look" to video. But movie production has gone digital. The production process is called digital intermediary (DI).

Compare the special effects of 20 years ago with those of today. If the desired effect could actually be achieved, it took an army of artists, working film frame by frame to produce a special-effect clip; it was a very expensive and time-consuming process. In a DI workflow, the first step is transferring the film content, frame by frame, to digital video. This is done by a datacine machine, the descendant of a telecine machine.

Once the image is in digital form, modern software-based editing and image processing systems can do anything—yes, anything—with the image. Although the analog film-to-digital video process can be expensive and time consuming, the time and money saved and the freedom from artistic limitations make the overall DI process cost-effective.

Copying the film master to produce thousand of distribution copies is another expense that can be reduced with films in digital format. Movies can be transferred as digital films over commercial content distribution networks to digital theaters. By comparison, the network connection is cheaper than film.

The major expense for a digital cinema theater system is having sufficient disk memory to store the digital "film." Consider a 10-screen multiplex cinema with a movie for each screen. Don't forget the space needed for a safety, backup copy.

Presentation of digital movies, as with DTV, is perfect. Digital files don't wear out, so there aren't any film scratches or other deterioration over a viewing lifetime. The thousandth presentation is at the same quality as the first.

Digital cinema is still in its infancy. But the number of locations where digital projection systems have been installed is growing rapidly around the world.

Getting back to movies repurposed for television, VoD is particularly intriguing from a technology and personalization perspective. It is a great example of how small audiences over time can add up to huge numbers. Following in the footsteps of the video rental store, VoD has emerged as a viable business. It is a lot more convenient to point the remote at the DTV and search for a flick for tonight's entertainment than it is to go to the rental store, hope you find something you like, stand in line, watch it, and then have to bring it back by a certain day and time. Need I say more in favor of VoD?

The only fly in the ointment is bandwidth limitations over the delivery channel and the number of sessions the server can handle. If the audience grows too large, the delivery network will choke.

As expensive as it is to produce feature films, it also costs a small (or large) fortune to promote them. Many new films that are not deemed theater-worthy are released straight to DVD and VoD. This enables producers to recoup at least some of their production costs. It is also an outlet for experimental, niche, or unusual films to be presented to a mass potential audience. Who knows—it is feasible that one of these could become a major hit.

The technology war over consumer recording to disk of HDTV content was over almost before it began. Supporters of the HD-DVD standard capitulated, and Blu-ray was left with a lock on high-definition consumer DVD technology. Blu-ray HD capable players are dropping to below $300, within the reach of most consumers.

Downloading full-length films over the Internet is becoming more practical now that network speeds and computer memory and processing power have increased dramatically. And although HD-quality content is still a little difficult to deliver, as compression codecs become more efficient and Internet speed increases, it won't be too long until HD and Internet distribution complete the fulfillment of the design requirement demanded of the HDTV candidate systems to be compatible with the information highway.

Digital cinema is here. Digital production, distribution, and presentation have advantages that the film industry cannot overlook.

The debt

It is important to always remember those who paved the way. The tendency is to forget that radio, black-and-white television, the camera, and movies were the result of the efforts of dedicated inventors.

We think of the homogenization of our contemporary media experience as amazing and revolutionary. And it is, for our moment in time. But 50 years from now, the world will wonder what it was like to get along without these innovations. They may never understand that a television and PC were different devices, or that a cell phone could not access the Internet.

So I take this brief interlude to acknowledge the efforts of the technologists, engineers, and scientists whose work brings us to the multiplatform universe we live in today. We all owe them a large debt of gratitude.

And now, back to our regularly scheduled programming.

TECHNOLOGY COORDINATION

Proprietary technology environments limit interoperability between consumer devices. This worked well for the deployment of the telegraph and telephone via government-sanctioned monopoly, and was used by IBM and Microsoft to lock anyone who bought their products into their entire catalog. IBM went so far as to develop its own digital representation of numbers and letters, EBCDIC, just to render ASCII useless on their computers.

Although some companies continue to try to leverage a closed, proprietary approach, new business models are evolving that use open standards in an effort to foster technology interoperability.

As with all technology, the establishment of standards can help insure device interoperability. Nowhere is this more important than in entertainment electronics, where consumers expect to turn on a DTV or any other device and transfer content with ease. No difficult user interfaces, or manual format conversions, or worse, no reboots or blue screens of death allowed!

Point of consumption interoperability

In an ideal world, transferring content between devices should be a simple as dragging and dropping a file into a folder. Any conversions necessary should be performed in a manner that is transparent to the user.

Digital content files usually include the most popular formats and standards. This includes WMA (Windows Media Audio), JPEG (photos), DivX (DVD using MPEG1 and 2, Dolby Digital & DTS among others), WMV (Windows Media Video), Flash Video, H.264, and MP3.

In addition to format and standards compatibility, there are medium format issues. Not all players can read content from all physical formats. A DVD/CD player serves as an example. Available players can access content physically stored in numerous formats: DVD-Video, CD-DA (digital "Red Book" audio), CD-RW (read/write), and CD-R (write once). A CD + R will not be readable by the player.

Another problem is backward compatibility. Older players cannot play MP3 files; others cannot display JPEG photos. In the future, as new codecs come into commercial use, today's standalone media players cannot be upgraded to support the new codecs.

Users who have made a computer the center of their media universe will (most likely) not have this problem. Software applications will appear on the market that will convert just about any media format to any other media format. Leveraging the plug-in capabilities of modern software, the conversion capability will be incorporated into the user's favorite media applications.

Consumer electronics association

Looking out for the consumer's interoperability interests is the Consumer Electronics Association (CEA), comprised of 2,200 companies in the consumer technology industry.

CEA originally came into existence as the Radio Manufacturers Association (RMA) in 1924. After a series of name changes, it was called the Electronic Industries Association (EIA) from 1957 to 1998, when it became the Electronic Industries Alliance (EIA). Old timers in the U.S. TV technology business sometimes refer to NTSC I as RS-170, which was the original black-and-white system.

In 1995, EIA's Consumer Electronics Group (CEG) became the Consumer Electronics Manufacturers Association (CEMA). Finally, in 1999, President Gary Shapiro announced the trade group's name change from CEMA to the CEA, and the organization became an independent sector of the EIA. In 2007, the EIA ceased to exist.

CEA produces the annual International Consumer Electronics Show in Las Vegas, the world's largest consumer technology tradeshow. Attendance has reached 140,000 retail buyers, distributors, manufacturers, market analysts, importers, exporters, and reporters from around the world.

The CEA is also a standards-setting organization. Often these specifications are set in conjunction with other industry groups such as the Society of Motion Picture and Television Engineers (SMPTE), the ATSC and the Audio Engineering Society (AES). The goal of standards-setting activity is to enable interoperability between new consumer products hitting the market and existing devices. The CEA Technology & Standards program is accredited by the American National Standards Institute (ANSI) and consists of over 70 committees, subcommittees, and working groups.

THE PRODUCTION RIDDLE

With so many devices to produce content for—each with its own capabilities—production infrastructure, processes, and workflows must be designed to maximize efficiency. This creative process is just a part of the overall media life cycle. In order for content to be presented on a device to the consumer, it not only has to be created, it has to be assembled, distributed, received, decoded, and presented.

Adding to the challenge is the fact that distribution channel bandwidth places constraints on audio and video formats. Just as Archimedes said, "give me a place to stand on, and I will move the earth," so too can anything be delivered, given enough bandwidth. But bandwidth is not unlimited, however, and each channel has a limit on the amount of data it is able to deliver in a given interval of time.

Production, repurposing, and distribution technology

Engineering is about solving a problem under real-world constraints. These problems may be technological, financial, environmental, or temporal. As described in this chapter, presentation of audio and video on each of the three-screen categories requires planning on all three areas. But it is totally dependent on technology.

Technology has enabled the unified media universe. All the required technological tools and capabilities are now available to support the three-screen media experience. Just a few years ago they were not. Before moving on to production systems and distribution channel methodologies, a discussion of the fundamental technologies that form the foundations of the modern digital media industry is necessary.

Media, technology, and the law

3

An argument can be made that the telegraph and then the telephone defined electronic communications; they were the keys that unlocked all other communications channels to come. The rollout of these technologies was of interest to the U.S. government, and thus the stage was set for regulation of the telecommunications industry. Individual countries the world over also regulate their communications industries, while the United Nations regulates international telecommunications.

Discovering how this government involvement came to be is helpful in understanding the influence government regulation will have on the proliferation of new content delivery channels. Many technologies have been touched by government regulation and policy, while many others have been shaped by lawsuits over intellectual property rights. Without a doubt, this will continue. Every media merger is scrutinized, while lobbying efforts seek to influence changes in policy or law that can be leveraged into profit.

Anyone who's been in the technology industry for any period of time eventually realizes that two factors determine if a breakthrough that makes it out of R&D will turn out to be a product that has a positive impact on a profit and loss statement: consumer acceptance and government regulation. Everyone in the media industry realizes that the best-laid plans can often be scuttled by either of these variable factors.

Today, the financial success of new business ventures and models in the media industry depends in large part on the resolution of issues such as establishment of limits on print and broadcast ownership in the same market; whether or not satellite radio is a national service; local carriage of terrestrial content by cable and satellite systems; and even the rights of copyright holders. These questions and others are usually resolved by government pressure and/or regulation. No company is exempt from the policies and decisions made by government agencies, bureaus, legislatures, and courts.

COMMUNICATIONS MEDIA OWNERSHIP

In some countries, the government owns the media. In the United States the media is run by the private sector and protected by the First Amendment to the U.S. Constitution as freedom of the press. But being a business, it is also subject to laws governing monopolistic practices. And since the early days when radio broadcasting began facilitated by use of the public airwaves, legislative bodies have established regulatory agencies to govern how the electromagnetic spectrum and other public resources are used for communication. Many times, the courts have been called upon to resolve differences of opinion.

In the beginning, AT&T created...

AT&T and its famed R&D unit, Bell Laboratories, has been a pivotal player from the beginning in the development of many forms of communications technology. Indeed, without AT&T and researchers at Bell Labs, modern communications systems may never have arrived at the sophisticated levels that we are accustomed to today.

In the United States, monopolies and unfair business practices are forbidden under the Sherman Antitrust Act of 1890, a bill that was sponsored by Senator John Sherman, the younger brother of General William Tecumseh Sherman of Civil War notoriety. However, because of the importance of the national deployment of communications systems, an exception was made in the deployment of telegraph and telephone systems, based on what was known as the Kingsbury Commitment.

The Kingsbury Commitment

In 1907, AT&T President Theodore Vail began acquiring companies that had business associations with Bell and other, independent systems. His acquisitions also included the purchase of Western Union. At the same time, AT&T would not allow independent systems to connect with their long distance lines. Together, these actions effectively eliminated all competition and created a virtual monopoly in the telegraph and telephone business.

The Attorney General's office was not pleased and pressured AT&T with legal action. In a 1913 letter from AT&T Vice President Nathan Kingsbury, AT&T agreed to divest itself of Western Union and to allow local and independent carriers and exchanges access to their long distance services. Any acquisitions were to be reviewed and approved by the Interstate Commerce Commission. This appeased the government for a while.

The U.S. Department of Justice (DoJ) began another investigation of AT&T in the 1930s to determine if they were in compliance with antitrust law. As a result, the antitrust suit *United States v. Western Electric, Co.* was filed in 1949. The action sought to separate Western Electric, AT&T's manufacturing division, from the company's telephone operations, breaking up the vertical integration structure.

While the action was in the courts, Bell Labs launched the Essex research project to develop computer-controlled transistor signal switching. This marked a movement into the fledgling computer industry and was outside the scope of AT&T's traditional telecommunications business.

In 1956 the suit was resolved when AT&T entered into a consent decree and agreed to limit their business interests to telephones and transmitting information, with the exception of military projects. Bell Laboratories and Western Electric would not enter fields such as computers and business machines. In return, the Bell System was left intact.

Not done yet

In 1974, the U.S. government began an antitrust suit against AT&T that was settled in January 1982. This time, AT&T agreed to divest itself of the wholly owned Bell operating companies that provided local exchange service, resulting in seven regional Bell operating companies (RBOCs). In return, the DoJ agreed to lift the constraints of the 1956 decree, allowing AT&T to enter the information service market. Divestiture took place on January 1, 1984, marking the end of the Bell System.

COMMUNICATIONS OVERSIGHT

The Federal Communications Commission (FCC), established by the Communications Act of 1934, is an independent U.S. government agency that is charged with regulating interstate and international communications by radio, television, wire, satellite, and cable. The FCC's jurisdiction covers the 50 states, the District of Columbia, and U.S. possessions.

One area of FCC interest has been to investigate and regulate the degree to which broadcast networks operate as monopolies. Consequently, FCC actions have had a large impact on the activities of broadcast networks. One must keep in mind that U.S. commerce policy is to stimulate marketplace competition.

Broadcast network diversity

GE created the Radio Corporation of America (RCA) in October 1919 to control Guglielmo Marconi's American patents. It was also responsible for marketing the radio equipment produced by GE and Westinghouse.

In 1926 RCA, GE, and Westinghouse bought radio station WEAF in New York. The National Broadcasting Company (NBC) radio network went on the air with 24 affiliated stations on November 15, 1926. The Rose Bowl game of 1927 was heard coast to coast, thanks to the NBC network.

In 1927 NBC's radio properties were divided in two. The NBC "Red" network offered entertainment and music programming while the NBC "Blue" network carried many of the "sustaining" or nonsponsored broadcasts, especially news and cultural programs. The colors may have originated from pushpins that engineers used to designate affiliates of WEAF (red pins) and WJZ (blue pins).

Beginning in 1938 the FCC investigated the radio industry and in 1940 issued a "Report on Chain Broadcasting." Finding that two corporate owners (and the cooperatively owned Mutual Broadcasting System) dominated American broadcasting, this report proposed "divorcement," requiring the sale by RCA of one of its chains.

In 1939 the FCC ordered RCA to divest itself of one of the two networks. Naturally, RCA fought the divestiture order, but it divided NBC into two companies in 1940 anyway, preparing for the eventuality of losing an appeal. The Blue network became the "NBC Blue Network, Inc." and the NBC Red network became the "NBC Red Network, Inc."

In actuality, the FCC did not regulate or license networks but issued a ruling that "no license shall be issued to a standard broadcast station affiliated with a network which maintains more than one network." NBC sued but the FCC won on appeal, and NBC opted to sell NBC Blue.

The sale closed on October 12, 1943. The new network, known as "The Blue Network," was owned by the American Broadcasting System, a company Edward Noble formed for the deal. In mid-1944, the network was renamed the American Broadcasting Company.

Media ownership regulation

Periodic reviews of media ownership regulations evaluate the diversity of information sources in a local market. News programming is of particular concern. It is often said that an informed public is necessary for the democratic process to function effectively. Hence, a variety of independent news sources are desired, so much so that it is the subject of legislation and debate.

Every four years, the FCC produces the Quadrennial Review Order. The report contains sections that address media ownership issues. These include cross-ownership of broadcast and newspaper properties, limits on network proliferation, and the number of stations that can be owned in a designated market area (DMA). The rules strive to implement what is in the public's best interest.

Rules adopted in the Quadrennial Review Order

The excerpts in the box are from the FCC Web site.

Newspaper/Broadcasts Cross-Ownership Rules

[W]hen a daily newspaper seeks to combine with a radio station in a top 20 designated market area (DMA), or when a daily newspaper seeks to combine with a television station in a top 20 DMA and (1) the television station is not ranked among the top four stations in the DMA and (2) at least eight independent "major media voices" remain in the DMA. A "major media voice" includes full-power commercial and noncommercial television stations and major newspapers. The Commission concluded that such sources are generally the most important and relevant outlets for news and information in local markets.

For markets smaller than the top 20 DMAs, reverse the negative presumption in two limited circumstances: when the proposed combination involves a failed or failing station or newspaper, or when the combination results in a new source of a significant amount of local news in a market.

No matter which presumption applies, the Commission's analysis of the following four factors will inform its review of a proposed combination:

1. the extent to which cross-ownership will serve to increase the amount of local news disseminated through the affected media outlets in the combination;
2. whether each affected media outlet in the combination will exercise its own independent news judgment;
3. the level of concentration in the DMA; and
4. the financial condition of the newspaper or broadcast station, and if the newspaper or broadcast station is in financial distress, the owner's commitment to invest significantly in newsroom operations.

Local television ownership limit

A single entity may own two television stations in the same local market if
   the so-called "Grade B" contours of the stations do not overlap; or
   at least one of the stations in the combination is not ranked among the top four stations in terms of audience share and at least eight independently owned and operating commercial or noncommercial full-power broadcast television stations would remain in the market after the combination.

Local radio ownership limit

The caps are based on a sliding scale that increases with the size of the local market. As a general rule, one entity may own

(a) up to five commercial radio stations, not more than three of which are in the same service (i.e., AM or FM), in a market with 14 or fewer radio stations;
(b) up to six commercial radio stations, not more than four of which are in the same service, in a market with between 15 and 29 radio stations;
(c) up to seven commercial radio stations, not more than four of which are in the same service, in a radio market with between 30 and 44 radio stations; and
(d) up to eight commercial radio stations, not more than five of which are in the same service, in a radio market with 45 or more radio stations.

The national television ownership limit

In 2004, Congress enacted legislation that permits a single entity to own any number of television stations on a nationwide basis as long as the station group collectively reaches no more than 39% of the national TV audience.

Radio/television cross-ownership limit

One company may own in a single market: one TV station (two TV stations if permitted by the local TV ownership rule) and one radio station regardless of total market size; or if at least 10 independent media voices (i.e., broadcast facilities owned by different entities) would remain after the merger, up to two TV stations and up to four radio stations; or if at least 20 independently owned media voices would remain post-merger, up to two TV stations and up to six radio stations or one TV station and up to seven radio stations. Parties must also comply with the local radio ownership rule and the local TV ownership rule.

Dual network ban

Retained the dual network ban. That rule permits common ownership of multiple broadcast networks but prohibits a merger of the "top four" networks, i.e., ABC, CBS, Fox, and NBC.

Multiple challenges to the Quadrennial Review Order currently are pending in the U.S. Court of Appeals for the Ninth Circuit, which will decide which court will ultimately hear these challenges.

(Source: http://www.fcc.gov/ownership/)

As you can see, the government takes an active interest in ensuring that the American electorate is kept informed by a number of media sources.

In other countries the level of control varies. In the United Kingdom the BBC is run by the government. It is done with a minimum of manipulation of the subject matter that makes it to air. As the world witnessed during the 2008 Summer Olympics, the Chinese government actively manages information that is disseminated over the broadcast media. It has even gone so far as to limit the information that enters the country over the Internet.

Media ownership regulation can be expected to continue as new media outlets increase their share of the content consumption audience.

TECHNOLOGY REGULATION

All too often marketplace dominance comes with a price. As has been discussed, national governments have a history of antitrust action against major technology companies.

Big problems for Big Blue

IBM had its first run-in with the law because of its dominance of early electronic accounting machines that processed data on keypunch cards. The company perfected a proprietary technology business model; once clients bought an IBM system they were locked into IBM. Connectors, cables, and even the code words used for characters were unique to IBM systems. Interoperability with another vendor's equipment was, for all intents and purposes, impossible. As one would expect, in time the government took notice and filed a suit in 1952.

Round 1

In 1956 IBM entered into a consent decree with the DoJ. The legal action was targeted at attaining two goals: promoting competition by creating a viable market for used IBM machines, and opening the market for electronic equipment used by companies that provided data-processing services such as payroll and bookkeeping.

To satisfy the first condition, IBM was required to sell rather than lease equipment. The court also demanded that IBM provide parts and information to companies that supplied independent maintenance for data-processing equipment.

To comply with the second condition, IBM was required to sell data-processing services through an independent business unit. As a result, IBM created a separate division, today known as the Integrated Systems Solutions Corporation.

Round 2

IBM's success in the mid-1960s again led to questions about possible IBM antitrust violations. The DoJ filed the complaint *U.S. v. IBM* in 1969. The suit alleged that IBM violated Section 2 of the Sherman Act by monopolizing or attempting to monopolize the general-purpose electronic digital computer system market, specifically computers designed primarily for business.

An action that has had long-term industry-wide consequences was the decision in 1969 to "unbundle" software from hardware sales. Until this time, software and software services were provided to purchasers of a vendor's hardware at no additional charge.

After the unbundling, IBM software was divided into two main categories: system control programming (SCP), which remained free to customers; and program products (PP), which were subject to a separate cost. Many industry analysts feel that this helped enable the creation of a software industry and opened the door for the establishment of independent computing services companies.

The DoJ persisted, and in 1973 IBM was found to have created a monopoly via its 1956 patent-sharing agreement with Sperry Rand in the decision of *Honeywell v. Sperry Rand*. This invalidated the patent on Electronic Numerical Integrator And Computer (ENIAC), a general-purpose machine.

IBM asked to be released from its 1956 consent decree. In 1995 the request was granted. A condition that IBM must continue to provide spares and manuals to third-party maintenance service providers and system brokers at a fair price was left standing. IBM would like to see the decree completely rescinded "as a matter of principle."

Microsoft 2000

No company is too large for the DoJ to set its sights on. Microsoft's troubles began in 1994. In order to avoid litigation, Microsoft entered into a consent decree to avoid accusations that it was engaged in "predatory" business practices in its dealings with computer makers.

Microsoft became the target of another antitrust and anticompetition action in 2000. The proposed solution was to break up Microsoft by dividing its operating system (OS) (Windows, NT, etc.) from the Office suite (Word, Excel, etc.)—another kind of unbundling, although this time it was to separate the OS from the applications.

At the time Microsoft controlled over 80% of the OS market and 90% of the market in business applications. Naturally, the company denied having acted anticompetitively, and pointed out that their success led to both standardization and relatively low-cost software.

Yet Microsoft's competitors, such as Netscape and Sun Microsystems, claimed that in blocking competition, Microsoft effectively stifled innovation. To this day, both technology and media companies continue to play the innovation card when convenient, as we'll soon see.

Microsoft avoided government litigation by entering into a consent decree in 2001. The provisions called for Microsoft to do the following:

- Allow end users and original equipment manufacturers (OEMs) to enable or remove access to certain Windows components or competing software (e.g., Internet browsers, media players, instant messaging clients, e-mail clients) and designate a competing product to be invoked in place of that Microsoft software.

- Disclose the internal Windows interfaces that are called by "Microsoft Middleware," which is separately distributed to update Windows. These interfaces are made available for third parties to use solely to interoperate with Windows.

- Make available on reasonable and nondiscriminatory terms the protocols implemented in certain Windows desktop OS products and used to interoperate or communicate natively with Microsoft server OS products.

However, Microsoft enjoyed the profits of a global presence. In 2004, after a five-year investigation, the European Union concluded that Microsoft had violated competition law by leveraging its near monopoly in the market for PC operating systems onto the markets for work group server OS and for media players.

Microsoft was also required to:

- Disclose to competitors the interfaces required for their products to be able to "talk" with the ubiquitous Windows OS.

- Offer a version of its Windows OS without Windows Media Player to PC manufacturers or when selling directly to end users.

Microsoft was also levied a $613 million (497 million euro) fine.

INTELLECTUAL PROPERTY AND THE LAW

The media industry is about technology and content. This makes it heavily dependent on intellectual property. While there have been many intellectual property and copyright cases over the years, several are especially illustrative of the impact copyright can have upon both creative and business endeavors.

My sweet infringement

George Harrison's 1970 hit "My Sweet Lord" was similar enough to The Chiffons' 1963 song "He's So Fine" to ignite a legal battle over copyright. Harrison explained that he was inspired to write "My Sweet Lord" after hearing the Edwin Hawkins Singers' "Oh Happy Day."

In the U.S. federal court decision in the case, known as *Bright Tunes Music v. Harrisongs Music*, Harrison was found to have committed subconscious plagiarism. The Copyright Act does not require a showing of "intent to infringe" to support a finding of infringement.

Harrison appealed, and the appellate court ruled that an infringement has occurred when the copyright owner demonstrates that the second work is substantially similar to the protected work and the second composer had "access" to the first work.

Harrison conceded that he had indeed heard "He's So Fine" when it was popular, thus establishing the second point.

The question at issue was that since Harrison's song was similar and created later than "He's So Fine" it was a derivative work, and therefore violated copyright law.

The slippery slope

Napster was born in 1999 of the desire to create an easier method of finding music on the Web than by using contemporary search engines. It became the first peer-to-peer file distribution system. Music was available as MP3 files and the enabling software application had a user-friendly GUI. The system facilitated the ability to download music from a large selection, available from other Napster users.

Naturally, the recording industry interpreted music "sharing" as equivalent to theft. Having come of age after the invention of the VCR, Napster users rationalized using the service for a number of reasons. It enabled them to freely obtain songs without having to buy an entire album. It was easy to download copies of songs that were hard to find. And of course the price for the songs was just right: free.

Individual users downloading songs might have been tolerated. However, some Napster users took advantage of the emerging ability of PCs to write data to CDs. Sometimes the CDs were distributed to friends. And not a dime went to the copyright holders in either instance.

Copying and distributing music is the bread and butter of the music industry. It didn't take long until the Recording Industry Association of America (RIAA) filed a lawsuit against Napster (December 1999). The basis of the suit was Napster's facilitation of the transfer of copyrighted material. Publicity around the trial spurred more people to use Napster. Users now numbered in the millions.

It got worse for Napster when a Metallica demo track became available before the commercial recording was released. The downloaded "bootleg" got radio airplay. And when the band discovered that a large portion of their catalog was also on Napster they filed a lawsuit. More legal problems hit Napster a month later. Rapper Dr. Dre sent a written request to Napster to remove his music; it was ignored, and he sued.

2000 was a really bad year for Napster. A&M records and several other recording companies also sued Napster based on provisions of the U.S. Digital Millennium Copyright Act (DMC Act).

The music industry claimed that:

- Napster's users were directly infringing the plaintiff's copyright;
- Napster was liable for contributory infringement of the plaintiff's copyright; and
- Napster was liable for vicarious infringement of the plaintiff's copyright.

The court found Napster guilty on all three claims.

Napster's popularity peaked in February 2001 with 26.4 million users worldwide. An appeal was rejected in March and by July 28, Napster had shut itself down.

Viacom v. YouTube, 2007

Social networking is all the rage today. Consumers who grew up in a world of unrestricted VCR recording and sharing of TV programs see nothing wrong with uploading video from a TV show. But not everyone in the media industry feels the same way about unauthorized use of copyrighted content.

In 2007, Viacom took legal action against the video-sharing Web site YouTube and its parent company Google. The basis of the lawsuit is—what else—"massive intentional copyright infringement." Viacom is seeking more than $1 billion in damages.

The complaint contends that (as of the time of its filing) nearly 160,000 unauthorized clips of Viacom's entertainment programming had been available on YouTube and that these clips had been viewed more than 1.5 billion times.

Viacom also asked the court for an injunction to halt the alleged copyright infringement and stated

> *YouTube appropriates the value of creative content on a massive scale for YouTube's benefit without payment or license ... YouTube's brazen disregard of the intellectual-property laws fundamentally threatens not just plaintiffs but the economic underpinnings of one of the most important sectors of the United States economy.*

Google responded to Viacom's lawsuit by claiming that it had not infringed on copyrights. An added argument played the "technological innovation impediment" card, saying that the lawsuit threatened the viability of the YouTube video-sharing Web site as well as others like it. Google's reasoning was that "by seeking to make carriers and hosting providers liable for Internet communications, Viacom's complaint threatens the way hundreds of millions of people legitimately exchange information, news, entertainment, and political and artistic expression."

So if everybody does it, does that make it OK?

Makes you wonder what's next, the freedom of speech card?

In fact that's not too far fetched.

In the fall of 2006 the Digital Freedom Campaign was launched. Its mission: "a national effort to fight back against efforts by the big record labels and movie studios to ban new digital technologies."

Members of the coalition include the Consumer Electronics Association, Public Knowledge, Electronic Frontier Foundation, Media Access Project, Computer and Communications Industry Association (CCIA), and New America Foundation. The group also said that it seeks to "educate policy makers, innovators, parents, students and other consumers about the lawsuits and legislation that threaten to revoke individuals' rights to use digital technology."

Their argument is based on the 1984 Supreme Court VCR time shifted viewing decision, which passed by the slimmest of margins: a 5 to 4 vote.

George Harrison was guilty of subconscious plagiarism; Google argues that if everyone does it, it's OK to violate copyright law. This begs the question: will copyright laws be ignored for consumer violations but be rigorously applied to professionals?

LAW OR STANDARD?

Throughout this book references will be made to standards set by various organizations. These will include:

- SMPTE: Society of Moving Picture and Television Engineers
- EBU: European Broadcast Union
- ATSC: Advanced Television System Committee
- SCTE: Society of Cable Telecommunications Engineers
- AES: Audio Engineering Society
- CEA: Consumer Electronics Association
- ITU: International Telecommunication Union
- ISO: International Organization for Standardization

Discussions will point out how certain standards will enable various technologies to be implemented; enable interoperability of equipment manufactured by different vendors; and in general shape future deployments of new technology.

This comes with a caveat: unless backed by incorporation of a standard into a legislated rule, regulation, or law, standards are voluntary. No one is obligated to comply with any standard. Vendors can implement a standard in its entirety, partially, or not at all. They are free to develop competing proprietary technology.

Development of the digital television standard in the U.S. is a relevant example. The effort to come up with an advanced television (ATV) standard, it can be said, began in 1982 with the formation of the ATSC. In 1988, the FCC got into the act by establishing a formal procedure to vet competing prototype systems and arrive at a viable technology. Along the way, ATV morphed into HDTV and analog systems gave way to digital technology.

When the dust cleared, the ATSC had documented the final system, developed by an industry consortium, the Grand Alliance. This document was brought before the FCC for adoption as the terrestrial digital television transmission standard and incorporated into Title 47 of the FCC Rules and Regulations (R&Rs). This makes the standard a legal document. Technical parameters must be met, or there will be FCC action and fines.

But a funny thing happened along the way. The development of a DTV transmission standard was influenced by both the computer and the motion picture industries. One of the points of contention was the inclusion of "Table 3," a specification of 32 video formats. If these were incorporated into the FCC R&Rs, every device manufactured to receive DTV terrestrial broadcasts would have to support all 32 video formats.

Objection was so strong that the FCC specifically eliminated Table 3 from the adopted standard. The result is that any presentation format, mix of presentation formats, or even an heretofore unseen format is now permissible. 1080i and 720p have been employed and specified by industry standards. No one has come up with a rogue video presentation format.

But 1080i and 720p are merely industry conventions that are not legally binding.

Regulatory compliance and technology

With respect to technology and broadcast infrastructures, equipment must be installed that will enable compliance with FCC R&Rs. This began with the authority to allocate spectrum at a designated frequency to a station licensee and continues to the present in mandating compliance with closed captioning requirements.

Some FCC requirements are meant to benefit the public. The Emergency Alert System (EAS) has been in place since the 1950s. Broadcasters are required to install EAS equipment at their own expense. For the record, the EAS national network has never been activated, not even on 9/11. It has found an alternate purpose and is regularly used for weather warnings and alerts.

MPEG-2 video and AC-3 audio compression is required in the DTV R&Rs. If you are a terrestrial broadcaster, you have no choice but to install MPEG-2 and AC-3 equipment and use it. Failure to do so is breaking the law. The same holds true for displaying program rating bugs and carrying PSIP data in your DTV bitstream.

Standards compliance

Voluntary standards, at least in theory, are established for the benefit of the industry. Interoperability is one of their primary functions; rather than islands of proprietary equipment, connect any 1080i serial digital interface (SDI) signal into any SDI capable routing system and you are assured that the signal will be routed properly.

But there is a "dark side" to technology standardization. Frequently, a company has such a dominant product that their technology becomes a de facto standard. Eventually, this technology may become a formal standard. Standards don't come with a free pass. Users of the standard will have to pay a licensing fee to the patent holder. If you want to implement advanced video codec (AVC) or video codec 1 (VC-1), contact the MPEG Licensing Authority and bring your checkbook.

INTO THE VAST UNKNOWN

The point of these discussions is that broadcasting and the media industry are now venturing into unknown territory. Besides technology issues and marketplace dynamics, the influence and impact of government actions and intellectual property rights may determine the success or failure of any new or traditional media delivery platform in the future.

Broadcast operation centers in transition

Digital technology has exponentially increased the complexity of television broadcast operations. On the surface, the big-picture view of a facility infrastructure looks pretty much like it always did.

As shown in Figure 4.1, content enters the facility from a remote, originates from a studio, or is played back from a storage media. Graphics are produced, audio and video are mixed in a program control room, and commercials are inserted along with bugs and closed captions in the master control room. The program signal modulates a carrier wave and the program is broadcast. It's as simple as that! Riiiiiight…

Once upon a time, an engineer could completely understand all aspects of a broadcast facility. Devices were based on hardware. Just look at the circuit schematic, gather data about devices if necessary, trace the signal flow and you knew just about all you needed to know. Those days are gone.

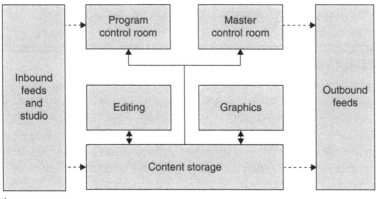

FIGURE 4.1

Functional block diagram of a broadcast facility

Today's broadcast infrastructures are too complex and too detailed for any one person to possess the requisite knowledge about every aspect of a system in sufficient detail to conceptualize, design, install, configure, and test a complete infrastructure.

If a broadcast engineer were to have to possess expertise not only in broadcast engineering, but also networking, computers, software, security and storage at a bare minimum.

Content that was once only stored on tape is now routinely stored on servers. Audio and video routing systems have been augmented with compressed domain IT network infrastructures. Nearly every piece of equipment has an RJ-45 connector and is connected to one of many networks. Paper rundowns have been replaced with newsroom computer systems (NCS). Even the lights, camera motion, and on-set displays are computer controlled. And these ain't your everyday home PCs and networks.

TRANSITION PAINS

This is a difficult time for veteran broadcast professionals. The proliferation of IT technologies in the Broadcast Operations Center (BOC) has forced veteran engineers to return to school. Nontechnical broadcast professionals are now confronted with a totally new technology and jargon. Careers based on sound engineering practices now have to be started all over again. Adapt or die!

Many an engineer's (and manager's) only experience with IT is via their PC. It is not unusual to find that freshmen technologists are more proficient in IT technologies than the experienced broadcast engineer can ever hope to be.

Even more dangerous is the tendency to marginalize, reassign to nontechnical roles, or even lay off veteran, traditional broadcast engineers. The philosophy is that IT streamlines workflows and processes and more can be done with less. Or for those of us old enough to remember, computers will save the world!

The Engineering–IT War

A side effect of the transition of television infrastructures to digital production and distribution precipitated a war in many broadcast organizations pitting Broadcast Engineering against Information Technology personnel. Internal technology departments often clashed over project turf. Lines of technology demarcation are blurred in multitechnology systems.

Prior to the digital transition, an organization would have had a broadcast infrastructure and a corporate computer network. E-mail and Internet access, back-office processing, and word processing was done on computers resident on the corporate network. Getting content to air was the function of the broadcast infrastructure. Never the twain did meet.

BOCs in this newly converged digital media infrastructure have a need for corporate and broadcast systems to interact. Commercial traffic, billing, and automation systems use much of the same data. Efforts by standards bodies have developed methods that enable back-office systems to communicate with each other and with broadcast equipment.

Some organizations have created departments that bridge the corporate network and broadcast infrastructure. In the digital media universe, broadcast and IT technologies have melded in an integrated infrastructure. By now there is a recognized need for broadcast media engineers who possess knowledge of both broadcasting and IT.

The war is over and neither side won.

A digital broadcast operations center

For most broadcasters their first HD broadcasts in late 1998 were either a feed through from a remote truck or a film to HD telecine transfer played back from a D-5 tape machine. The workflow required only HD master control and HD transmission capabilities: compression, transport stream multiplex, channel coding, and modulation. An HD production infrastructure was not required. Existing graphics, logos, and other elements were used sparingly and upgraded to HD when necessary.

Many parts of the production infrastructure were already standard definition digital capable. CCIR601 and SMPTE 259 had been around for over a decade and many infrastructures had incorporated SD-SDI signal distribution. The reduction in video cabling, from five cables (RGBHV) to a single coax cable, and improvement in video and audio quality alone made the upgrade to digital cost effective.

Transitioning to a digital infrastructure comes in two flavors, a "green field" start-from-scratch building of a completely new facility, or converting the current infrastructure to digital. If one has the luxury of constructing a new facility, budget and commissioning deadline will be limiting factors.

Converting an analog BOC to digital is challenging in that you must stay on the air while the transition is going on. In many facilities, portions of the infrastructure will have to be replaced in a step-by-step manner. When a new digital control room is built, air integrity can be protected by running both control rooms in parallel; if the new one fails, switch back to the old. If analog production and distribution resources have been replaced with new digital installations, in many instances the old analog infrastructure will be converted to digital.

Deceptively simple

Television broadcasting is expanding to multiplatform delivery and this complicates infrastructure technology even more than just going digital. What was once just television, now includes Web streaming, delivery to cell phones, pay per view (PPV), VOD, DVD, and other ways of repurposing the original program content. A technical infrastructure that facilitates this "any asset, in any format, instantly available, anywhere" is the Holy Grail requirement that is the guiding design philosophy in any transition to digital production.

This chapter abstracts broadcast operations into a conceptual framework that can be used to partition media systems into their component technologies. A familiarity with fundamental broadcast engineering and IT is assumed. The appendices contain short technical overviews of these topics. In this chapter, technologies and systems

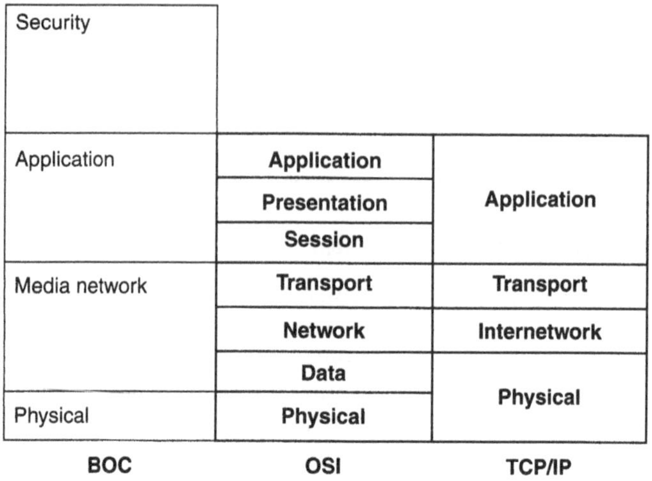

Security		
Application	**Application**	**Application**
	Presentation	
	Session	
Media network	**Transport**	**Transport**
	Network	**Internetwork**
	Data	**Physical**
Physical	**Physical**	
BOC	**OSI**	**TCP/IP**

FIGURE 4.2

Comparison of four-layer BOC (broadcast operations center) with OSI and TCP/IP models

that are of particular relevance to a BOC are discussed. In Chapter 8, workflows will be discussed with respect to the supporting infrastructures.

BROADCAST OPERATIONS CENTER TECHNOLOGY LAYERS

Broadcast engineers have always had to deal with maintaining a 100% reliable operation. Any equipment "hit," no matter how small, that impacts an on-air broadcast is totally unacceptable. Mastering the proper skill set and developing the expertise and attitude necessary to manage broadcast television technology can occupy an engineer's entire career.

Yet today's digital BOC has exponentially increased in complexity with the addition of IT technologies; each of these disciplines is challenging and career-consuming if one aspires to technical mastery.

A BOC can be conceptually modeled as consisting of four layers. As shown in Figure 4.2, these are the physical, media network, application, and security layers.

It is important to remember that this is just a model of an infrastructure. Its purpose is to help group and organize systems and resources in a way that facilitates a "big picture" understanding of the overall infrastructure.

If numerical values of complexity can be ascribed to each layer, the total BOC complexity has probably increased by a factor larger than a googolplex!

The physical layer

At the foundation of a modern broadcast infrastructure lies the physical layer, where the long-established technologies and practices of broadcast engineering have reigned

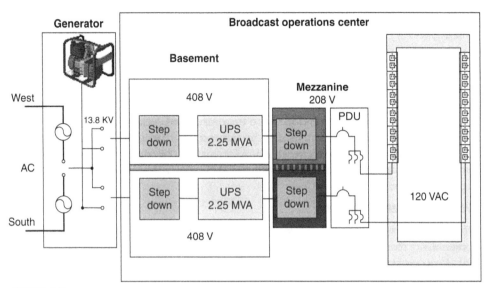

FIGURE 4.3

Redundant power distribution

since the birth of television. Systems required for production workflows that process, distribute, and assemble uncompressed audio, video, and graphics comprise the physical layer of a broadcast infrastructure. All fundamental infrastructure support for the entire building/infrastructure is also part of this layer. This includes electrical power, heating, ventilation, and air conditioning HVAC, and communications systems.

Power distribution

Analog and digital signals are susceptible to electro magnetic interference (EMI), radio frequency interference (RFI), and other forms of noise. Grounding systems adequate for analog systems will need to be upgraded. Digital systems are not as tolerant of voltage fluctuations and current spikes or dropouts.

Broadcast power systems are designed for reliability and uninterrupted operation. Figure 4.3 illustrates how the redundancy may be implemented.

Power enters the complex from two completely independent paths on the commercial electrical grid (left side illustration) from different geographical directions. Redundant generators and uninterrupted power supply (UPS) feed dual distribution circuits up to two power strips on the racks. Equipment is powered by dual power cables to independent power supplies.

Intercoms

Intercom (IFB) and all other production and support communications, telephones, and walkie-talkie radios must be installed, configured, and maintained. Communication while on air must be 100% reliable with appropriate backup. If any equipment malfunction or operator error threatens the integrity of the on-air signal, support must be reached instantaneously.

Real-time content formats

Uncompressed audio and video are real-time, continuous signals. Although digital production has brought many compression format choices, distribution and routing of baseband audio and video is done with uncompressed serial digital signals.

However, serial digital formats are for transport of baseband signals. There are a number of audio and video presentation formats to choose from. They are called presentation formats because this is what a DTV receiver will produce after decoding the audio and video.

Video presentation formats

Where there was once one analog National Television System Committee (NTSC) presentation format, now there are many: 1080i, 720p, and 480i. A digital video format is a combination of aspect-ratio, pixel-grid, scanning method, refresh rate, and color space. More information about each attribute is available in Appendix E.

Table 4.1 lists valid values for each format attribute.

As an example, 1080i in the ATSC system is the result of the following combination of attributes:

Aspect Ratio:	16×9
Pixel-Grid:	$1,920 \times 1,080$
Scanning:	Interlaced
Refresh Rate:	59.97
Color Space:	ITU 709, 4:2:0

Color space consists of two attributes: color gamut and sampling method.

Table 4.1 Global Digital Video Presentation Formats

Aspect ratio +	Pixel-grid +	Scanning +	Refresh rate (Hz) +	Color sampling =	Video format
16×9	1920×1080	Interlaced	60	4:4:4	1080i
4×3	1280×720	Progressive	59.94	4:2:2	720p
16×10	1440×720		50	4:2:0	VGA
14×9	720×576		30	3:1:1	QCIF
	720×480		29.97		
	1820×1200		25		
	1280×768		24		
	640×480		23.98		
	320×240				
	176×144				

Audio formats

Digital audio formats are somewhat different from video formats in specification. The attributes include the number of audio channels and types of services.

Audio channel format is denoted in terms of surround sound. For example, 1.0 designates mono, 2.0 stereo. 5.1 is five-channel surround sound with a low-frequency effects subwoofer. The ".1" designation is derived from the limited frequency response of the channel, which results in a reduced data rate, roughly one-tenth of a full channel.

Services are divided into main and auxiliary. Main services are complete main or music and effects; basically a mix—minus, everything but dialog. Auxiliary services include voice-over, dialog, and others.

An example of the power of audio services is the ability to release a program with 5.1 music and effects as the main service audio and add dialog in the appropriate language for the intended geographic location.

Graphics formats

The first difference between graphics and television video is color space. Television uses luminance and color difference signals while graphics systems use the RGB color space. Somewhere in the production chain, graphics will be converted to video color space.

This can create a problem and may produce illegal colors when graphics are combined with video. Besides the color space issue, computer graphics (CG) systems use the full 8-bit, 0–255 range whereas ASTC legal colors are in the 16–235 range. Therefore, graphics data must be scaled to the ASTC video range.

Quality control (QC) is extremely important. Conformance to legal color space must be verified. Errors can be corrected by color space/gamut legalizers, available from a number of broadcast equipment vendors.

Signal distribution

Because a wide variety of audio and video formats are available in a digital broadcast facility, due to analog legacy material and infrastructure, routing systems have evolved beyond their fundamental function of allowing real-time signals to be switched from any source to any destination. Conversion between analog and digital, embedding and de-embedding audio, up- and downconversion, transitions, and auto failover capabilities can now be incorporated into routing equipment, simplifying workflows, and reducing the need for discrete conversion equipment.

Baseband audio and video

Component analog video distribution requires three precisely timed distribution signals for each image component: red, green, and blue. Horizontal and vertical synchronization pulses are necessary for equipment to operate "frame" synchronously; each analog channel or AES pair of audio will require a cable.

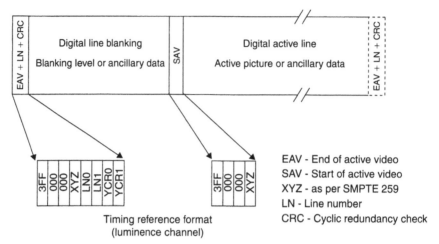

FIGURE 4.4

Serial Digital Interface (SDI) for SMPTE-292 (HD) and SMPTE 259 (SD) television lines

Analog signal distribution is practically nonexistent in a newly constructed facility. SDI is used almost exclusively. When analog signals from legacy equipment are used, conversion to SDI is done as close to the analog source as possible.

Using SDI rather than analog component video reduces the required discreet signal paths to one and will also eliminate the need for component timing verification and precise cable lengths. H and V sync signals are a part of the SDI signal. By embedding audio in an SDI signal, only one wire is necessary, greatly simplifying routing and distribution. Figure 4.4 shows the structure of an SDI signal.

The illustration represents for a single line of digital video. EAV and SAV are analogous to a sync signal. The other signals aid in validating data integrity. The only real difference between D SMPTE 259 and HD SMPTE are the number of samples in the active pixel area.

System evolution

Early SDI routing systems were designed for SD 601 serial digital signals. As HD-SDI found its way into broadcast operations, the first HD-capable routers only supported the 1.485 Gbps HD data rate; separate SD and HD distribution was required. Similarly, serial digital transport interface (SDTI), SMPTE 310, and asynchronous serial interface (ASI) routing were either not possible or required a dedicated routing infrastructure.

SDI speeds continue to increase (see Table 4.2). 1080p60 has spawned 3 Gbps standards. Vendors have addressed the 3 Gbps requirement by exploiting the modular design of their routing systems. Existing routers with backplane and connection schemes capable of supporting 3 Gbps bandwidth can be upgraded by replacing input, output, and cross-point circuit boards in existing frames.

However, a limitation to increased data rates is the existing cable infrastructure. Due to eye pattern degradation, existing coaxial cable run lengths may not support

Table 4.2 SDI Standards and Associated Data Rates

Standard	Name	Bitrates	Video Formats
SMPTE 259 M	SD-SDI	270 Mbit/s, 360 Mbit/s, 143 Mbit/s, and 177 Mbit/s	480i, 576i
SMPTE 292 M	HD-SDI	1.485 Gbit/s, and 1.485/1.001 Gbit/s	720p, 1080i
SMPTE 372 M	Dual Link HD-SDI	2.970 Gbit/s, and 2.970/1.001 Gbit/s	1080p
SMPTE 424 M	3G-SDI	2.970 Gbit/s, and 2.970/1.001 Gbit/s	1080p

3 Gbps data rates. This may create problems if a facility migrates to 1080p60-based production or desires to implement faster than real-time HD-SDI signal distribution.

1080p60 signals may be able to be distributed over existing cabling. HD-SDI capable cable lengths will be cut in half for 3 Gbps signals. Installation of re-clocking distribution amps may not be possible for existing cable runs. Additionally, 3 Gbps signals essentially are RF signals in the L and S bands where existing cable crimps, wire nicks, and tight bends can easily degrade a signal.

Blanking intervals and ancillary data

A video frame is constructed from SDI lines in a manner similar to an analog raster. The structure of a 1080i video frame is shown in Figure 4.5.

The blanking intervals are of particular interest and are referred to as ancillary data areas. Every line has a horizontal blanking area that in an SDI signal is called the horizontal ancillary data area (HANC). The vertical blanking area is referred to in a similar fashion as the vertical ancillary data space (VANC).

HANC and VANC are used to carry information. The structure of an ancillary data packet is specified in SMPTE 291. Its structure is illustrated in Figure 4.6.

Data IDs (DID) signal the type of data that is carried in the packet. SMPTE xxx specifics the value of a DID for each type of data. Data count, payload, and checksum complete the packet.

Audio

Audio distribution can be either embedded, discreet, time-division multiplex (TDM), or Dolby E. If audio is embedded, it will follow the video, on a single cable, but must be de-embedded for any processing. This will require embedding/de-embedding equipment appropriately placed in the signal distribution path. TDM uses a single wire and must be encoded and decoded. Dolby E uses lossless compression to move multiple channels of audio on wire pairs. Discreet audio will require an individual wire for each signal, significantly adding to the audio router port count.

The common denominator for audio is the AES/EBU specification of the AES3 format.

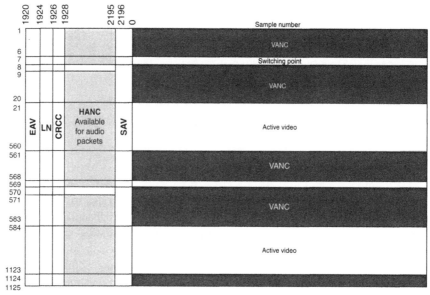

FIGURE 4.5

VANC and HANC ancillary data areas

SMPTE 291:
    Ancillary Data Packet and Space Formating
SMPTE 334:
    Vetical Ancillary Data Mapping for Bit-Serial Interface

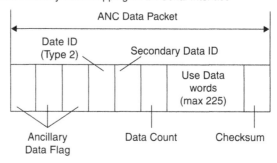

FIGURE 4.6

ANC packet format and data mapping

An AES3 signal consists of a stream of PCM audio samples; usually a pair. Figure 4.7 illustrates the structure.

More than a single AES3 pair can be carried over an AES3 distribution or routing system using one of two different techniques.

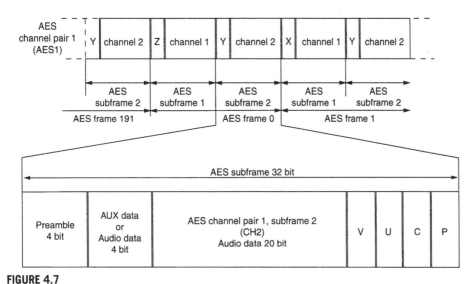

FIGURE 4.7

AES3 digital audio stream frame structures

Multichannel audio digital interface (MADI) is documented in AES10-2003. The MADI standard includes a bit-level description. MADI is similar to AES3 but as its name implies, is not limited to an audio pair. 28, 56, or 64 channels, sampling rates of up to 96 kHz and word lengths of up to 24 bits are supported.

MADI links use a transmission format that is similar to the FDDI networking technology (ISO 9314) and MADI has been used in talent micing and production intercomm systems.

The other method, Dolby E, uses light compression to squeeze multiple channels of digital audio into an AES3 compatible serial digital signal. Dolby E will be discussed in a later chapter.

Audio and video

The most common use of the HANC area is to embed AES3 audio. Figure 4.8 shows the format as specified in SMPTE 299.

Embedders and dembedders must be installed at appropriate points in the signal and workflow. But the benefit of moving audio and video around the BOC as a single signal can offset the need for additional equipment.

Sync distribution

In a broadcast plant, all switching operations occur synchronously during the video vertical interval blanking. A single house sync format will be used to genlock all incoming and outgoing feeds by passing the signals through a frame sync. This will most often be an analog NTSC black burst or black signal. Tri-level HD sync pulses have not been widely implemented in broadcast equipment.

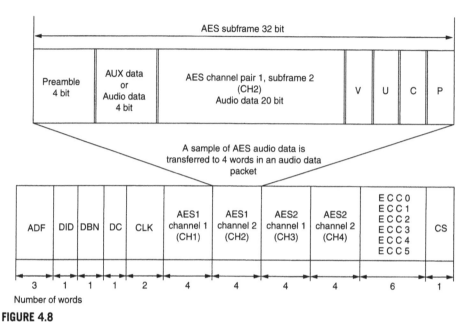

FIGURE 4.8

AES audio can be embedded in SDI HANC areas as specified in SMPTE 299

Production and master control switching is timed to the vertical blanking interval. Time code is distributed as part of sync system.

However, with multiple formats of video being used in production, it is common to have a sync signal for each format available from a routing and distribution infrastructure. Audio sync signals including digital audio reference signal (DARS) and word clock are available.

Sync is so important that redundant generators with auto failover capabilities are nearly always part of the system design. Time of day is synchronized by using GPS receivers.

Routing systems

Taking sources to air requires seamless, real-time switching of baseband audio and video. No viewer or advertiser will ever accept long periods of black between segments, or fragments of clips or commercials on the head and tail of program segments. Problems with commercial insertion and delivery can cost broadcasters significant sums in make-goods. A problem during a Super Bowl broadcast can impact the revenue stream by millions of dollars.

Facility routing systems consist of input and output connections, an X–Y switching matrix, and control panels. Audio, video, and control signals are connected from an input to one or more outputs. Digital SDI, AES, and control data (over RS422) are dominant signal formats and routers are referred to as audio, video, or data routers. Although rapidly disappearing, analog audio and video are still present in many

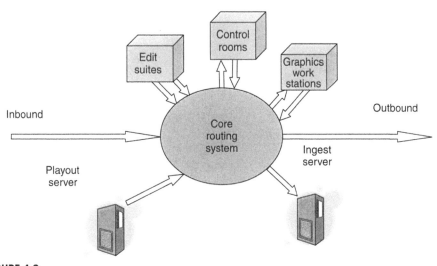

FIGURE 4.9

Centralized routing system architecture

facilities. Mono and stereo audio, composite (CVBS composite video baseband signal) and component (RGB or occasionally YUV) along with horizontal and vertical sync signals are found in legacy content.

To address legacy and the myriad ever-changing format interface requirements, router systems utilize a modular design on the circuit board level. I/O (input/output) cards may include the capability to convert between analog and digital formats, embed and embed audio and video, transparently handle 270 mbps, 1.5 Gbps, and 3 Gbps SDI signals. Many of the latest generation of interface cards will accept fiber optic cables.

Control panel interconnections are migrating to IP network technology. Otherwise, the RS422 serial data interface is used.

Routing systems are synchronized, referred to as "genlocked," to the house reference signal. This can be black, black burst, or tri-level sync. Switching occurs during the vertical blanking interval, usually as specified in SMPTE RP 168.

A salvo is a technique that enables the router control to be configured to make more connections between inputs and outputs in a single, sequential operation. These custom, facility-dependent sequences must be programmed by the vendor, system integrator, or in-house technical personnel.

Routing topology

Maximizing signal availability is desired in broadcast operations. This has led to a distribution infrastructure where one large (often huge) house router is fed every source and feeds every destination. Frequently, the physical router is partitioned into physically dispersed frames, with redundant signal paths. In this way, if one portion of the router fails, signals can be routed through a secondary path, ensuring uninterrupted operations (see Figure 4.9).

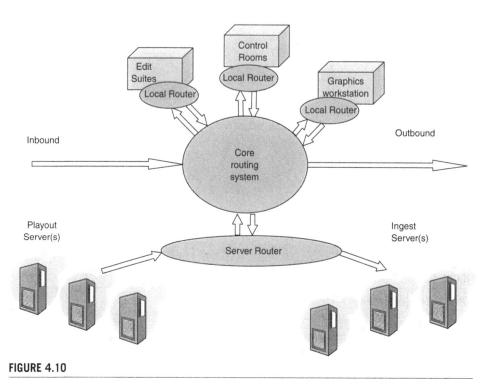

FIGURE 4.10

Spoke and hub, local and centralized routing architecture

There is a trend to augment the centralized router with small local routers that serve control rooms, QC, ingest, edit, and graphic areas, as shown in Figure 4.10. Local control panels are configured with limited source and destination cross-point control.

A complete centralized routing system is illustrated in Figure 4.11. The relationship of the components is simple in concept. But designing a routing system involves great care. I/O ports will become scarce; mnemonics used in routing tables must be descriptive. Signal access rights must be applied to routing control; you don't want to have an editor accidentally reroute a signal going to air.

Devices that are an integral part of digital production are the ingest and playout servers. They are in the lower right of the illustration. They accept SDI and AES connections to the routing system. They are also the door to the next layer in this model: the media network.

Routing compressed content over facility routing systems

Broadcasters have made huge investments in their routing and distribution infrastructure. Existing systems took years to install; personnel are comfortable using the control panels. Everyone is familiar with signal locations and destinations.

SDTI and ASI were developed as methods of routing MPEG-2 transport streams over 270 Mbps SDI routing systems. A brief explanation of SDTI follows: the reader is urged to pursue more information about ASI if he/she so desires.

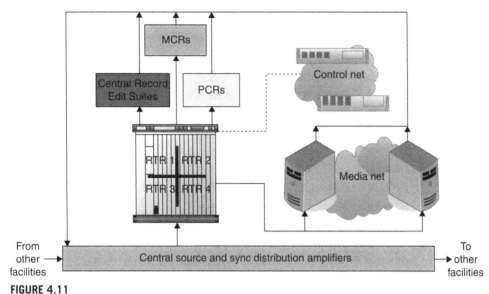

FIGURE 4.11

A functionally complete routing system with a partitioned router, control network, and servers

Data transport using SDI

SMPTE 305 specifies a method for using an SDI format for the conveyance of pure digital data. It solves the inherent problem of data that is all "0"s or all "1." This can be a problem because the SDI format identifies the start (SAV) and end (EAV) of active video by a sequence of bytes (FF, 00, 00) containing either all "0"s or all "1"s.

As Figure 4.12 illustrates, 8 bit data words are converted to 10 bits by repeating bit 7 for bit 8 of the new 10 bit word, and then inverting this bit for bit 9. This simple technique makes it impossible for any data word to conflict with SAV or EAV.

SDTI has found application in live, remote events. Using 270 Mbps or 1.5 Gbps SDTI fast turn-around replays in particular are well suited to the technology (see Figure 4.13). The systems are disk based and content can be accessed in a nonlinear fashion; scrubbing through a tape is a thing of the past.

The End of Traditional Broadcast Engineering?

The infrastructure described in this section is the scope of traditional broadcast engineering. Reliability and redundancy are key; these systems must never break down and if they do, switching to a redundant system of signal must happen transparently and instantly.

Even though these systems are complicated and it takes a career in the broadcast industry to attain technical mastery of them, a broadcast engineer could gain expertise in all necessary systems and technology.

8B → 10B	8B → 10B	8B → 10B	
B0 = B0	B0 = 0 B0 = 0	B0 = 1 B0 = 1	In this way 8-bit data with a ×00 or ×FF value can never exist in an STDI stream
B1 = B1	B1 = 0 B1 = 0	B1 = 1 B1 = 1	
B2 = B2	B2 = 0 B2 = 0	B2 = 1 B2 = 1	
B3 = B3	B3 = 0 B3 = 0	B3 = 1 B3 = 1	
B4 = B4	B4 = 0 B4 = 0	B4 = 1 B4 = 1	Therefore. SAV and EAV codes will only occur in the correct positions.
B5 = B5	B5 = 0 B5 = 0	B5 = 1 B5 = 1	
B6 = B6	B6 = 0 B6 = 0	B6 = 1 B6 = 1	
B7 = B7	B7 = 0 B7 = 0	B7 = 1 B7 = 1	
B7 = B8	B7 = 0 B8 = 0	B7 = 1 B8 = 1	
$\overline{B7}$ = B9	B7 = 0 B9 = 1	B7 = 1 B9 = 0	
	0×00 0×200	0×FF 0×1FF	

FIGURE 4.12

8B/10B coding converts 8-bit bytes into 10 words by copying B7 to B8 and then creating B9 by inverting B7

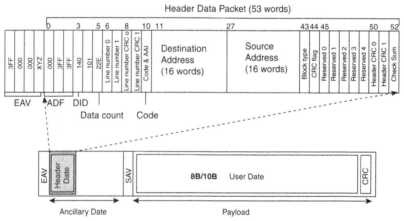

FIGURE 4.13

Serial digital transport interface format (Derived from SMPTE 305M)

As will be discussed next, the "infiltration" of IT, computer science, software-based systems, applications, and security have changed the playing field.

THE MEDIA NETWORK LAYER

Moving up the technology stack from the physical layer, the domain of traditional broadcast engineering, to the next layer in the four-layer broadcast infrastructure model, we enter the Information Technology realm. The media network, application, and security layers are the dominion of computer science.

With the digitization of sights and sounds, the transition to digital production has created the need to efficiently distribute "essence" in a way that better suits digital data. Network transfers of uncompressed video are categorically out of question in a BOC; SDI/AES routes can do that. But by employing audio and video compression, content files can now be distributed around the broadcast infrastructure on a "media network" using existing IT network technologies. This can result in efficiency improvements in many processes.

Ingest and playout servers are the portals to the media network from the physical layer. Ingest is the process of compressing baseband signals, generally SDI, into an MPEG, digital video (DV), or other production format file, attaching descriptions (metadata), and logging the content in an asset management system. Playout is the converse operation, and uncompresses, or decodes, essence files for real-time, baseband presentation.

With the use of compression technology, the traditional baseband facility routing infrastructure has been augmented by the addition of a new dimension: file-based audio and video routing over a media network. In some workflows, content may never exist in the uncompressed domain while in the facility. The integration of networks and broadcast routing systems has extended the scope of a facility routing and distribution infrastructure; media networks must be considered part of the audio and video routing system. This should be taken into account during conceptual design.

On the media network layer, files are the principal unit of information transfer through the production process. These files are broken down to packets for transfer depending on the routing protocols used.

A media network can be thought of as consisting of two elements: the routing resources that comprise the network and the media storage devices. This chapter will focus primarily on the network.

Audio and video file sizes are routinely measured in gigabytes and total system capacity storage in terabytes and petabytes. Disk-based storage is integrated with automated tape libraries to implement hierarchical storage management systems. Content backup strategies must be developed based on risk and access need.

It is important to understand that a BOC media network is neither a home PC network nor an IT network; think secure mission-critical military, avionics, or medical scenarios where a system failure can result in loss of life. However, although no one's life depends upon them, content file transfers must happen reliably and in a timely manner. Anything less can result in the loss of revenue.

The interconnection components of a media network are the Cat 5 and 6 cables that connect switches, routers, and bridges.

Compression for production

Video formats such as MPEG, DV, AVC, and VC-1, along with AES, MADI, or AC-3 and Dolby E digital audio streams, may have to be supported. This broaches the issue of transparent file interoperability and transcoding artifacts.

Table 4.3 Popular Production Compression Codecs

Codec	Video bit rate (Mbps)	Bit depth	Compression	Standard
HDCAM	135	8	DCT based (intra)	SMPTE 367M–368M
DVCPRO HD	100	8	DV based (intra)	SMPTE 370M–371M
HDCAM-SR	440	10	MPEG-4 SP (intra)	SMPTE 409-2005
XDCAM	35, 50	8	MPEG-2 (GoP)	–
DN × HD	36, 145, 220	8, 10	DCT based (intra)	SMPTE VC-3
Infinity	50–100	10	Wavelet based (intra)	JPEG2000
AVC-I	54, 111	10	AVC (intra)	High 10 intra profile

Video compression

Numerous production codes have been developed for use during production. Table 4.3 lists a few of the attributes of three different format families: DNxHD, DV, and HDCAM.

Video compression is extremely complicated and a detailed discussion is beyond the scope of this book. However, some of the more popular codecs will be covered in greater detail in Chapter 9.

Dolby E audio compression

Dolby E is a technology that can fit 10 audio channels in an AES pair and can be transported in the HANC area of an SDI signal. The primary benefit is that this package of multiple audio channels is compatible with an AES routing system.

As Figure 4.14 shows, the encoding process produces a packet that, from a format perspective, is identical to AES3.

Maintaining Audio and Video Sync

A side effect of the compression process is the difficulty of maintaining synchronization between audio and video. This is due to the fact that compression encoding and decoding delays vary for audio and video. Somewhere during the production workflow they must be realigned.

Computers, formats, and graphics

Graphics format interoperability is particularly challenging in that graphic application vendors have built islands of technology. File formats are proprietary and cannot directly be used by other graphics systems. For example, a Deko file cannot be edited on an AVI system.

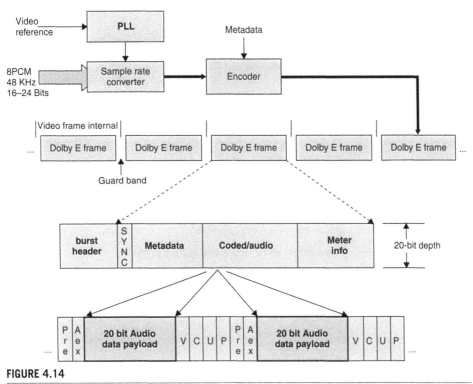

FIGURE 4.14

The relationship between Dolby E and AES audio

Some vendors have addressed this by designing equipment that converts file formats on the fly transparently in the normal production workflow. No extra manual conversion step is necessary. Conversion can be accomplished by dragging and dropping files to be converted into a watch folder. The conversion application will convert the content and move it to a defined location, all without any additional manual intervention.

Types of broadcast infrastructure networks

Nearly every new piece of equipment in a broadcast infrastructure has an RJ45 connector on it. Everything must be connected to a network. This integration of networking capabilities into broadcast technology has enabled new methods of remote device configuration, control, and monitoring.

In most network topologies, devices are connected to the nearest switch. The physical connections are divided logically into network topologies. A typical division of networks by subsystems is listed in Table 4.4.

Each broadcast subsystem network should be as independent and redundant as possible. A common IT practice of connecting networked devices to the nearest switch or router, and then configuring subsystems in logical networks does not

Table 4.4 Broadcast Networks

Network	Function examples
Routing control	Connects routing system control panels for audio, video, RS-422 and time code
Audio file server	Used by audio production studios: to produce and mix music and dialog for segments and shows
Broadcast media	All high-resolution audio and video is transferred between ingest, MAM, and playout servers
Graphics	Supports production of graphics, animation, stills, headshots, and special effects
KVM	Keyboard, video, and mouse switching interconnection for computers and computer-based equipment
Asset management	PCs used to catalog, store, and retrieve content
Production systems	Play-to-air graphics PCs, playlist PC, newsroom computer system
Multiviewer display	Controls PCR, MCR, and studio monitor walls and displays
Corporate network	E-mail, Internet
Automation systems	Computer-based systems that control ingest and playout devices
Studio systems	Show control, cameras

provide the physical redundancy necessary in a broadcast environment. Creating a "flat" network with functional distinctions based on subnets is dangerous. A broadcast engineering philosophy of reliable, robust, and resilient operation must be applied. Intelligent physical network separation is a must.

As Figure 4.15 shows, in the broadcast world, that redundancy is the rule.

Frame Size

The size of a packet has a direct impact on network speed. Each packet transfer has an associated overhead. An Ethernet packet can be up to about 1,500 bytes. A Jumbo Frame can be around 9,000 bytes. Six Ethernet frames are contained in one Jumbo frame. This facilitates fast file transfers due to the fact that only one sixth of the overhead is necessary for a Jumbo Frame transfer as compared to an Ethernet transfer. This can result in large efficiency gains when file sizes are large, as with HD content.

Analysis of production workflows is necessary to determine the optimum network topology. The challenge is to establish deterministic worst-case file-transfer speeds.

For example, we are all familiar with the phenomena of indeterminate network performance. One day an Internet site can be lightning fast, the next, slower than

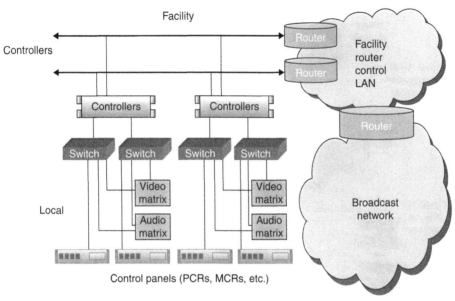

FIGURE 4.15

Two control panels are located in a PCR, MCR, etc. and control the signals needed for that operation. Users with proper authorization can access any resource on any portion of the routing system over the router control LAN. Full system redundancy is implemented

a glacier. To the average person accessing a Web site, this is merely annoying. In a broadcast infrastructure, unpredictable network performance is unacceptable.

Using routing protocols with quality of service (QoS) traffic engineering capability will enable defining guaranteed minimum acceptable file-transfer rates over the various subnetworks.

Storage

Media storage technologies are the second component of the Media Network, and deserve at least a word or two. Many different kinds of storage technologies such as fiber channel (FC), RAID, storage area networks (SAN), direct attached storage (DAS), and digital tape are now an integral part of a broadcast infrastructure.

Storage selection has become more complicated with the development of network attached storage (NAS), SAN, and DAS options. Careful system design choices must be made to determine the best mix of storage technologies, performance, and cost. Figure 4.16 shows a variety of storage types in a typical graphics system.

IP address assignment

Designing the network and numbering scheme requires careful thought and planning. Every device connected to any network, be it broadcast, production, or

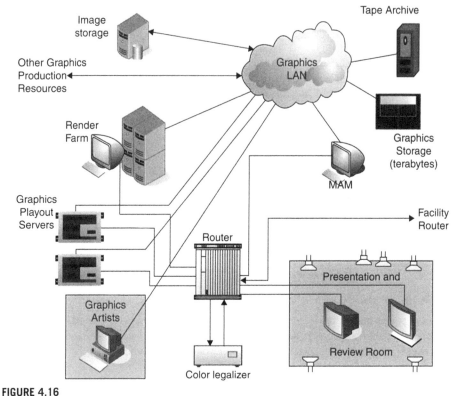

FIGURE 4.16

Generic use of storage in a graphics production infrastructure

corporate, will require an IP address. While MAC addresses are fixed in firmware, IP addresses will have to be assigned.

Two fundamental methods of assigning IP addresses are available: static and dynamic. Static addresses are permanent and live with the device even if power is recycled. Dynamic addresses are assigned only when a device connects to the network and attempts to communicate with another networked device.

Static addresses

Assignment of static IP addresses is a completely manual operation. A network architect will develop, at least, the overall network topology before considering the IP assignment scheme. It is a top-down approach.

In a well-designed network, physical and logical topologies, namespace design, IP address assignment, and security configuration are intimately connected. To blindly assign IP addresses may create problems later with security, sign-on, and network scaling. If hundreds of devices have been configured with static IP addresses and the subnet needs to be expanded to include more devices, altering the existing IP scheme may prove to be impossible.

Dynamic host configuration protocol

Not every device needs to be connected to a network all the time. Use of static IP devices, in this scenario, would be highly inefficient.

The dynamic host configuration protocol (DHCP), IETF RFCs 1541 and 2131, automates the assignment of IP addresses, subnet masks, default gateway, and other IP parameters to a client computer device from a DHCP server. This simplifies administration of a large IP network

A DHCP server manages a pool of IP addresses and information about client configuration parameters such as the default gateway, the domain name, the DNS servers, and other servers such as NTP time servers. A DHCP lease is the amount of time that the DHCP server grants to the DHCP client permission to use a particular IP address. A typical server allows its administrator to set the lease time.

DHCP defines a number of methods for allocating IP addresses:

- Dynamic: assigns leased IP addresses. Periods range from hours to months. A client will use the renewal mechanism to maintain the same IP address throughout its connection to a single network.

- Automatic (DHCP reservation): the address is permanently assigned to a client.

- Static allocation: allocates an IP address based on a table with MAC address/IP address pairs, which are manually filled in. Only requesting clients with a MAC address listed in this table will be allocated an IP address.

- Manual: the address is selected by the client and the DHCP protocol messages are used to inform the server that the address has been allocated.

The automatic and manual methods are generally used when maximum control over IP address is required.

The dynamic process of address allocation is known as ROSA: request, offer, send, accept. Figure 4.17 illustrates the communication sequence.

When a DHCP client connects to a network:

1. The client computer sends a broadcast request (called a DISCOVER or DHCPDISCOVER), looking for a DHCP server to answer.

2. Network routing directs the DISCOVER packet to the DHCP server.

3. The server receives the DISCOVER packet and determines an appropriate address to give to the client.

4. The server then reserves that address for the client and sends an OFFER (or DHCPOFFER) packet back to the client with that address information.

5. The server also configures the client's DNS servers, WINS servers, NTP servers, and other services as necessary.

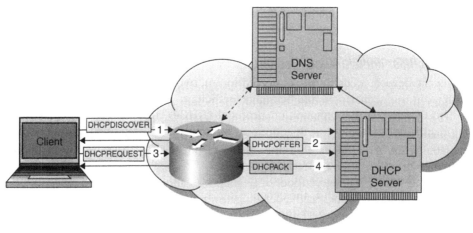

FIGURE 4.17

DHCP protocol communication flow. Devices are assigned IP addresses from an available pool as they connect to a network

6. The client sends a REQUEST (or DHCPREQUEST) packet, letting the server know that it intends to use the address.

7. The server sends an ACK (or DHCPACK) packet, confirming that the client has been given a lease on the address for a server-specified period of time.

Once the REQUEST packet is received, the DHCP server will assign an IP address, a lease, the subnet mask, the default gateway, and other network configuration information.

The DISCOVER process is often initiated immediately after powering up the device. As soon as the DHCP assignment and configuration process is completed the device can communicate with other devices on the network.

CIDR route summarization

Classless inter-domain routing (CIDR) is a replacement for the managing networks based on Class A, B, and C addresses. It is also called supernetting or route summarization.

Figure 4.18 illustrates the technique.

In the example, consider the graphics network. The network address is the first three octets. Each subnetwork would require an entry in routing tables. CIDR enables access to these networks by using the common portion of the addresses, the first two octets from the aggregation router. This minimizes the number of network addresses in the aggregate router's routing table. In a large network, this will have a big impact on minimizing packet latency.

Network discovery

Networks are dynamic. Devices are connected and disconnected often. Keeping track of the configuration and all the devices on a network can be overwhelming.

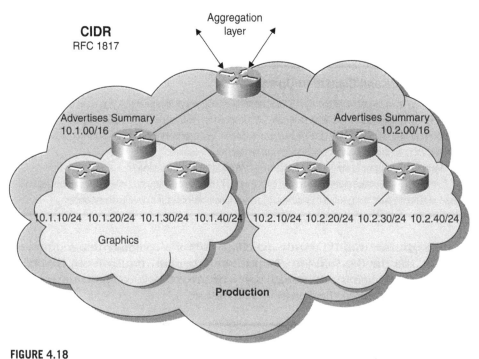

FIGURE 4.18

Route summarization enables access to subnet resources via a single routing table IP address entry

Keeping track of this information has been simplified by an automated technique called network discovery.

An application runs a series of tests by sending special packets over the network. These are specified in Internet connection management protocol (ICMP), IETF RFC 792. Devices respond, IP addresses are documented, and a network topology map is drawn.

THE APPLICATION LAYER

Modern broadcasting could not function without computers. From stand-alone graphics PCs to multiterabyte archives, computers, and the applications that run on them are an integral part of any broadcast facility.

Computers can be embedded in equipment or stand-alone as multipurpose work-stations. NCS, media asset management system (MAM), playout automation, and graphic creation tools are some of the application-layer systems that facilitate program production and dissemination.

Computer platforms are the interface between the application layer and the media network. Performance, stability, and interoperability of these computer-based systems must meet broadcast operational requirements.

The operational requirements of application-layer resources can be better under-stood if broken down into logical groupings; computer platforms, Operating Systems

(OSs), and applications. Platform performance must be sufficient and reliable. OSs must be stable. Applications must be interoperable.

Deterministic and Consistent Operation

Nondeterministic performance is unacceptable. It is very annoying to have an application that runs as fast as lightning one day and then for no apparent reason shows snail-like performance the next day. But because modern OSs are very complex and run many processes in the background consistent performance is elusive.

Memory leaks get blamed for everything. What is a memory leak?

Similarly, networks have the same problem. By their very design, layer 3 IP networks will use different paths to deliver "packets," small divisions of a file, over the network.

Although built with PC boards and chips, 90% of a computer lies in firmware, software, and the OS. Software and platform OSs must run flawlessly. Having to restart an application is bad enough, but having to reboot a computer or recycling power is not an option when a station is on the air.

Open Source versus License Fees

There is a debate about open source versus vendor proprietary OSs and software. The benefits of open source are no license fees and access to source code. On the downside, with the exception of the Internet user community, you are on your own regarding support. Therefore many choose to license an OS and purchase support. But this can be expensive.

A time may come when using open-source resources becomes cost effective to the extent that redundant systems can be installed to improve reliability. The other approach is to pay for support with the implied result that the system will be more reliable.

Graphics systems

The use of innovative graphics is an important part of a creative process. Great graphics can really make for great programming. But without a sufficiently engineered and reliable supporting infrastructure, graphics creation will be stressful and frustrating.

The beginning and end of every graphics (GFX) production workflow is the GFX workstation.

Typical minimum system requirements for a GFX workstation:

- Processor(s): quad or eight core/processor, 64 bit, $>3.0\,\text{GHz}$, 800 MHz FSB, hyperthreaded

- RAM: 4 GB 400 MHz DDR

- System Hard Disk: 7,200 RPM (minimum)

- Media Hard Disk: high performance RAID

- NIC: GigE

- Display: the native resolution of the video format (1,920 × 1,200 or higher)

The internal computer architecture is important. This will ultimately determine application performance. Chipset compatibility with the applications intended to be used should be verified. It is a given that there are separate system and media drives running on discreet controllers.

Important features of any GFX card are SD/HD capability, OS compatibility (Windows or MAC OS X), HD tri-level sync, and RS-422 machine control. To lay HD renders and animations to tape will require a consistent throughput of 200 Mbps (that's bytes) to accommodate the 1.5 Gbps uncompressed stream.

Applications and reliability

An operator's portal to the entire broadcast infrastructure is through the application layer. Be it GUI or command line, switches and LEDs, or LCD readouts on a front panel, somehow personnel must reliably command and control the broadcast infrastructure.

Application stability is affected by many factors. Does the machine have enough horsepower to execute the installed applications? Are users running other programs in the background, like listening to music or surfing the Internet?

Code can be compiled or interpreted. For compiled programming languages, source code is written and then compiled into an executable program in machine code. Interpreted languages process code line by line at run time into machine code that is then executed. Compiled is fast, interpreted is slow.

Interoperability

Interoperability is the capability for different systems to communicate with each other in a meaningful fashion.

Installing applications straight out of the box may work fine for isolated islands of production processes. But when using multiple software applications, authored by numerous vendors, interoperability problems will often arise. Work with the vendors to ensure that your installation meets your requirements and functions reliably.

Middleware

With the continual increase in OSs and application complexity and variety, it has become nearly impossible for an application to be written with complete platform portability. Interaction with system hardware is enabled by the use of drivers, little chunks of code that interface with system resources like disk drives and graphics cards. At the user interface, an API provides an application developer with methods of interfacing with the input and output channel and devices.

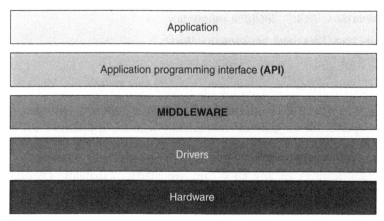

FIGURE 4.19

Middleware abstracts communication between hardware and applications enabling a software developer to write a defined application programming interface

The connection between the user interface and API and the Hardware drivers is accomplished by the intervention of middleware. Figure 4.19 illustrates the communication stack.

A programmer need only write to the API interface. The driver communicates with the hardware. Middleware handles communication between the drivers and API.

Application development

An organization with programmer talent may choose to develop custom applications. A benefit is that you will retain complete control of the source code, modifications, and patches. On the other hand, you will be responsible for modifications and patches.

Adequate testing must be done during development and after installation. This includes proving that an application does what you want, and figuring out what happens when something goes wrong. Will there be a "graceful degradation" of the application or will the computer lock up? What happens if a network cable is unplugged? It is important to know how a failure affects your workflow or on-air functionality, before something goes wrong!

BROADCAST INFRASTRUCTURE SECURITY

Think of the glory and fame that a "script kiddie" can claim if they can produce a "Nipplegate" incident on the air. Certainly this is motivation enough to try to hack a broadcast infrastructure.

That disgruntled employee who didn't get promoted might be tempted to get even with a little surprise insert while you are on the air. Do you want to be responsible for something far worse than the Super Bowl Nipplegate incident? How will you know who did it?

Nuclear power plants are secure. Right? A few years ago an article in the *Proceedings of the IEEE* told a tale of a nuclear power plant that was infected by a vendor's laptop. Fortunately, the reactor was down at that time.

Security attacks should be of critical concern to broadcasters. Malicious content tampering or an infrastructure breach can be embarrassing, damage business relationships, or take you completely off the air. Yet at all levels, from management to the production staff, broadcast personnel are painfully naive about broadcast infrastructure security.

Inadvertent acts by broadcast infrastructure personnel are also a security concern. Internet connectivity and transferring files from a nonsecure computer (from home) can infect a platform, the media network, and the entire infrastructure.

Security methods

Restricting physical access to buildings, rooms, and equipment has been the primary method of maintaining security at broadcast facilities. Show your badge and you're in. Is this still sufficient?

Physical access limitations are not sufficient. Virus scans and spam blockers used in a home network are inadequate for use in a BOC alone. Firewalls and access control lists (ACLs) will not fully secure a BOC. Passwords can be broken. How do you deploy intrusion detection systems (IDS) and intrusion prevention systems (IPS) so that they are most effective?

Layer by layer

The security layer consists of increasingly tightening layers of protection. This requires a hacker to repeatedly attack and breach protection mechanisms. You might get lucky and the hacker will give up. Designing for security must be a fundamental infrastructure requirement.

Controlling physical access is the first line of defense. This is generally done by using badges, card reading locks, and security guards, and by locating equipment in locked rooms. But with so much broadcast infrastructure equipment being on a network, a hacker can get into your facility without physical access.

Media network security consists of firewalls, proxy servers, demilitarized zones (DMZs), and ACLs. These techniques must be intelligently and cost effectively deployed and intimately integrated into the broadcast infrastructure to be effective.

Network security techniques

An understanding of widely used network access concepts and techniques will underscore the challenge in securing a broadcast infrastructure.

Firewalls

A firewall is a method where devices are configured to permit, deny, encrypt, decrypt, or proxy all computer traffic between different security domains based on a set of rules.

Classifications of firewalls is based on where the communication is taking place, where the communication is intercepted, and the state that is being traced.

Network layer and packet filters. Network layer firewalls, also called packet filters, operate at layer 2 and 3 of the OSI model and the Internetworking layer of the TCP/IP stack. Packets are allowed to pass through the firewall only when they match a defined set of rules. Network administrators define the rules or the default rules will apply. These depend on how they are configured by the vendor.

Stateful firewalls maintain context about active sessions and use that "state information" to process packets. If a packet does not fit the profile of earlier network traffic, it will be intercepted. On the other hand if it compares favorably with the firewall's state table, it will be allowed to pass.

Stateless firewalls require less memory, perform filtering rapidly, and are used to filter stateless network protocols. Because of their simplicity, they cannot perform sophisticated packet analysis and filtering.

Firewalls filter traffic based on packet attributes including source IP address, source port, destination IP address or port, and destination application such as WWW or FTP. They can filter based on routing protocols, time to live (TTL) values, domain name of the source, and other attributes.

Application layer. Application-layer firewalls work on layer 7 of the OSI model and layer 4 of the TCP/IP stack (i.e., all browser traffic, or all telnet or FTP traffic) and may intercept all packets traveling to or from an application.

Proxy server

A proxy server acts as an intermediary between a client computer and a network resource. For example, a computer wants to access a Web site. However, the network is configured such that all requests for Internet access must go through a proxy server. Since the Web page does not reside on the proxy, it will forward the request to the Web site. The response will also pass through the proxy server and then be passed on to the requesting computer.

Firewalls may be implemented by using a proxy server. If all requests and replies are passed unmodified, the proxy may be called a gateway or, sometimes, a tunneling proxy.

Demilitarized zone (DMZ). A DMZ is an area between the Internet and the facility network.

As Figure 4.20 shows, two firewalls separate the Internet, or an outside network from an interior network. The area between the two firewalls, the DMZ, is occupied by mail servers, Web servers, proxy servers, and IDS.

Access control list (ACL)

An ACL sets permissions, or access rights, to files, directories, resources, or networks.

File system ACLs are usually based on a user's ID or username. Access permissions control read, write, or execute capability. In a UNIX or LINUX environment, the command CHMOD sets permissions. Bit patterns set rights for a user, groups, or all.

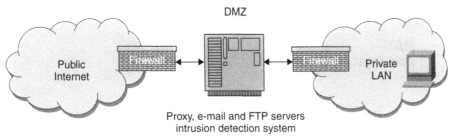

FIGURE 4.20

A Demilitarized Zone (DMZ) is an area that uses firewalls wall to separate networks and create an area where proxy servers and security devices can monitor traffic flow between the networks

```
Access-list {1-99 or 1300-199} {permit or deny} source-address
{wildcard mask}
Router(config) #access-list 1 deny 10.0.0.55
Router(config) #access-list 1 permit 10.0.0.0
ip access-group {number or name} {in or out}
Router(config) #interface ethernet 1
Router(config-if) #ip access-group 1 out
```

LISTING 4.1

A generic ACL configuration command set

When used in a network, ACLs allow or block network traffic based on IP source and destination IP addresses. Allowed and blocked addresses are listed during router configuration.

For example: a standard ACL filters on source address only as shown in Listing 4.1.

ACL entries are evaluated in order. Therefore, this list denies the computer at 10.0.0.55 access but allows all others on the 10.0.0.0/8 network.

The wildcard mask defines the address space that will be examined. Specified in four octet dotted notation, IP addresses are filtered based on "0"s and "1"s. Opposite in format from a subnet mask, "0"s determine address bits to filter on. For example, a wildcard mask of 0.255.255.255 matches all Class A networks.

An extended ACL can filter network traffic based on source address, destination address, and/or protocol. Additionally, if TCP or UDP is encapsulated, source and destination ports can be used as filtering triggers. ICMP (IETF RFC 792) messages can also be filtered.

Operators include equal, greater than, less than, and range.

Network address translation (NAT)

Originally developed to address the limited number of IPv4 routable addresses, NAT (IETF RFC 1631) uses public and private addresses in conjunction with internal and external IP addresses to hide the internal network address from the outside connections. Addresses in packet headers are modified by a routing device. The result is the remapping of one address space into another.

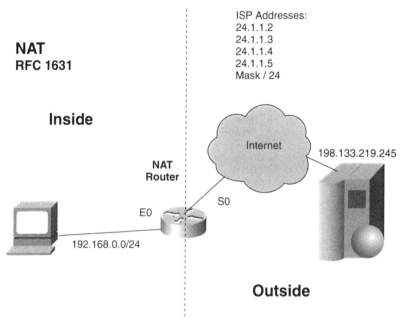

NAT
RFC 1631

ISP Addresses:
24.1.1.2
24.1.1.3
24.1.1.4
24.1.1.5
Mask / 24

Inside

Internet

198.133.219.245

NAT
Router

E0   S0

192.168.0.0/24

Outside

FIGURE 4.21

Network Address Translation (NAT) translates an IP address used by an inside, local network to an IP address used by an outside, global network

As shown in Figure 4.21, firewalls often have NAT functionality. Hosts protected behind a firewall have addresses in the "private address range," as defined in RFC 1918.

NAT can be implemented in a number of ways.

A static NAT maps an unregistered IP address to a registered IP address on a one-to-one basis. A computer with the IP address of 192.168.32.10 will always translate to 213.18.123.110. This is useful when a device needs to be accessible from outside the network.

A dynamic NAT maps an unregistered IP address to a registered IP address from a group of registered IP addresses. A computer with the IP address 192.168.32.10 will translate to the first available address in the range from 213.18.123.100 to 213.18.123.150.

Port address translation (PAT)

PAT, also referred to as overloading, maps multiple unregistered IP addresses to a single registered IP address by using different ports (see Figure 4.22).

NAT and PAT router configuration   Command sequences used to configure static and dynamic NAT and PAT capabilities are presented in the following listings:

```
Static NAT:
    Ip nat inside source static < inside local IP > < inside
    global IP >
    NAT(config) #ip nat inside source static 192.168.0.25 24.1.1.2
Dynamic NAT:
    ip nat pool [pool-name] [first-IP] [last-IP] netmask [mask]
    NAT(config) #ip nat pool MyPool 24.1.1.3 24.1.1.6 netmask
    255.255.255.0
    NAT(config) #access-list 1 permit 192.168.0.0 0.0.0.25
    NAT(config) #ip nat inside source list 1 pool MyPool
PAT:
    NAT(config) #access-list 1 permit 192.168.0.0 0.0.0.25
    NAT(config) #ip nat inside source list 1 interface serial 0
    overload
    In all three cases, the router configuration is completed with
    the following assignments to the router interfaces:
    NAT(config-if) #interface e0
    NAT(config-if) #ip nat inside
    NAT(config-if) #interface s0
    NAT(config-if) #ip nat outside
```

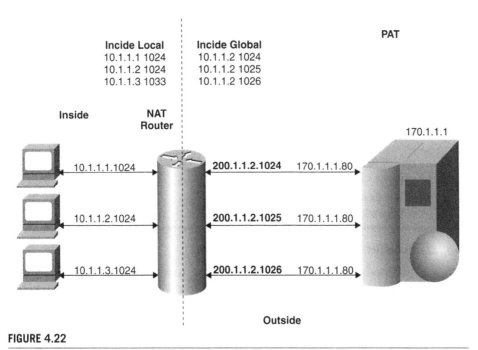

FIGURE 4.22

In port address translation (PAT), each computer on the private network is translated to the same IP address (200.1.1.2), but with a different port number assignment

Virtual private network (VPN)

A VPN is a method that creates a logically independent and secure route from the public Internet into the corporate network. It can be used to enable access to BOC resources, images, content from a remote venue, other stations, or from personnel working from home.

There are two types of VPN: remote access and site-to-site.

Remote-access VPN. A remote-access VPN, also called a virtual private dial-up network (VPDN), is a user-to-(local area network) LAN connection between the private network and remote locations. Remote-access VPNs permit secure, encrypted connections between a private network and remote users through the Internet.

An example is journalists in the field sending their latest segments back to the BOC over the Internet. Another is distributing content to remote broadcast sites.

Site-to-site VPN. By using dedicated equipment and encryption, you can connect multiple fixed sites over the Internet.

Site-to-site VPNs can be:

- Intranet-based—Using a corporate network, remote locations can be joined in a single private network. In this way, the graphics LAN in LA could be securely connected with the NY graphics LAN.

- Extranet-based—Enable connecting diverse networks together. For example, the graphics intranet VPN may need to include a connection to a creative house. Limited resources on the vendor's LAN could be connected (with limited access rights) to the graphics VPN.

Never the Twain Shall Meet

To connect or not to connect, that is the question
   Whether it is safer in the quest for content transfer
   To limit access to the chosen few
   Or permit a touch point with the great Internet cloud
   And run the risk of suffering
   The slings and arrows of potential attacks
   And look like an idiot to all your colleagues
   And risk the wrath of the FCC

The role of active directory and single sign on

On the application layer, granting of access rights to accounts is based on username and password. By using a "least privilege" approach, a graphics artist can only log onto a graphics workstation, not a playout server.

Risk assessment

Security attacks can originate externally and internally. Security problems can involve employees or people who have gained physical access, such as visitors. Actions can be accidental or malicious.

Incidents have two phases. In the attack phase, a way into the infrastructure is sought. When a way in is found, there is a breach.

A thorough security audit, a common practice in the IT world, will examine all these issues. This is an assessment of threats that have the potential to cause harm. An effort is made to identify areas of vulnerability. A determination of the impact of an incident will guide implementation of security techniques and technologies.

Basic methods

Security should be an integral part of infrastructure design. There should be no one point of vulnerability.

Network segregation

It's a good idea to segregate on-air broadcast networks, production networks, and corporate networks. In this way, only the corporate network touches the Internet. This places the obstacle of at least two additional layers of security between an attacker and on-air systems. Figure 4.23 shows a conceptual network architecture.

Holes in the machine

Computer platforms are vulnerable in a number of ways. Graphics artists and other personnel frequently work at home. Content is copied onto a USB stick or DVD and transported home and back to the facility. The home computer is probably not secure. Inadvertent infection of the production machine and potentially all devices on the work can happen.

Lockdown is the act of removing or disabling services and applications on a computer system that will not be used and are not necessary. This removes potential security holes.

Virus scanning should be installed on every computer. Periodic checks of entire machines and on-access scanning of a file as it is being opened should be implemented. Virus definition files need to be updated weekly.

However, security has a price. It can be expensive, time consuming, and require dedicated personnel. Every level of security slows infrastructure speed. Evaluate the impact of security implementation on infrastructure performance.

A Response Plan: Limiting the Damage

Every facility needs to have a plan about how to respond to an attack. Once you've been breached it is too late to figure out what to do.

Establish an incident response team (IRT) to respond to attacks, investigate incidents, make evaluations, and initiate corrective action. An IRT consists of technical specialists, corporate security, legal, HR, and executive management among others. From this group, a core team is established to perform threat and risk analysis, vulnerability testing, and security audits. Appropriate response plans are developed for any possible incident. Security alerts are published.

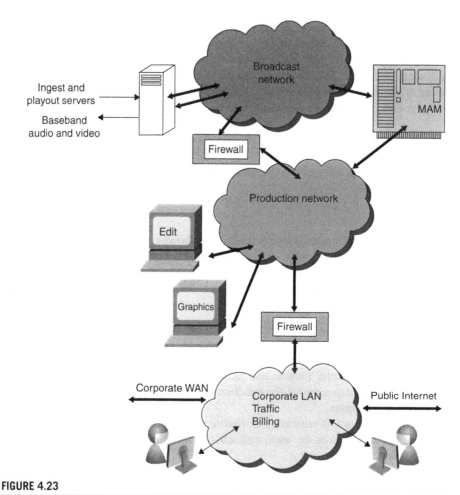

FIGURE 4.23

Organization network functional separation

An IRT operation process is defined. A consistently applied approach must be followed. Incidents must be documented and a database created. If there is an incident, an analysis will identify measures to take to prevent future attacks.

If an incident occurs, intrusion response personnel must be alerted instantaneously. A sufficient amount of information must be gathered so that an audit trail can be traced and the responsible party identified.

Continuous monitoring

Implementing real-time monitoring capabilities that immediately make security personnel aware of a potential attack is imperative (see Figure 4.24). IPS and IDS are

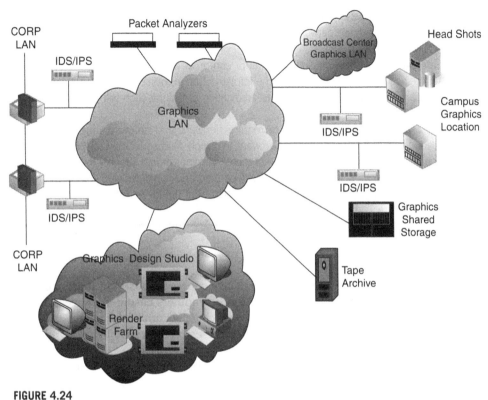

FIGURE 4.24

Network security is greatly enhanced by the strategic location of Intrusion Protection (IPS) and Intrusion Detection (IDS) systems

fundamental security techniques that activate monitoring system alarms and notify the IRT.

IPS will block a suspected attack before it can enter the network infrastructure. These are previously seen and analyzed attacks. An IPS must be used in conjunction with other techniques. Hackers are clever.

IDS monitor networks, hosts (OS), and applications. Some technologies analyze activity patterns. If suspicious activity is observed, actions are taken to prevent the suspect file or activity from inflicting any damage and to issue alerts.

Internet Access and Security

With the infusion of IT into the broadcast environment comes the threat of a breach of infrastructure security. Computer, application, and network policies that are adequate for most businesses are not even close to safe for broadcasters.

Consider the hazards of Internet connectivity. Active code, routinely downloaded from the Web, can include viruses, worms, and bots. Virus scans, intrusion detection, and prevention are not always effective against zero-day attacks. Dual homed computers that connect to the broadcast network and the Internet through separate network interface cards (NICs) can still be hacked.

Internet worms, viruses, and other forms of computer attacks are a threat to knock a BOC off the air. They can travel around the infrastructure in milliseconds. Every networked resource can become infected. Complete eradication of the infection will at least be very time consuming or at worst impossible.

Some broadcast equipment vendors require firmware and software upgrades to be downloaded over the Internet. They may want to supply field service by a VPN into your machine and take control remotely. Each of these scenarios entails security risk.

Broadcast networks should be physically separated and totally isolated from corporate networks and the Internet. As a stand-alone system, the broadcast infrastructure can be considered secure.

A matter of trust

A broadcast infrastructure may not be a military command post, but there are security techniques to learn from the DoD. Trusted computers ensure file traceability by using a unique username and password. OSs include security information as data is input to system over a secure network. When these conditions are met, OSs and networks are certified.

Similar protocols belong to a networked broadcast infrastructure. Test and certify computer platforms and OSs before installation.

Knowledge is power

Broadcast personnel need to become knowledgeable about security techniques, technologies, and infrastructure deployment. Train all levels of personnel to be security aware. Security is a philosophy.

- Use a combination of physical and logical measures.

- Each user should have a unique account. Reset passwords.

- Implement the "least privilege" concept.

- Require users to read and sign off on the security policy when their accounts are created.

- Display a "usage" notification when users sign on prohibiting unauthorized use and stating that violators will be prosecuted. Employees should not expect privacy.

- Enforce security policies.

- Warn and discipline employees when necessary.

- Perform security audits regularly.

■ Never let a user log on with administrator privileges!

Senior management must understand the need for security.

Be afraid. Hacking is becoming an organized, criminal activity. Access to your infrastructure can be sold to groups that want to do damage to your organization. And without adequate security, you will not be aware that you are at risk. Be VERY afraid!

SYSTEM MONITORING

Comprehensive BOC infrastructure monitoring using a single application is the Holy Grail and presently is unattained. However, there is a solution that can be implemented by using monitoring systems that currently exist for each of the four BOC layers.

Monitoring BOC infrastructure resource health is neither a luxury nor a trivial task. With the integration of broadcast and IT technologies, along with the resultant exponential increase in infrastructure complexity, neglecting to implement centralized monitoring methodologies will result in confusion and stress.

Among the challenges of monitoring infrastructure health are:

■ The larger the facility, the more physically disperse the equipment

■ Support personnel with experience in broadcast and IT technologies are rare

■ Integration of discreet monitoring systems under one GUI

These issues make it essential that equipment be centrally monitored so that when failures occur, they can be swiftly localized, clear instructions are given to "first responders," and all relevant information for each subsystem is available for analysis.

A VERTICAL TECHNOLOGY STACK

This chapter has presented a conceptual framework that models integrated technology broadcast systems as a vertical stack. An operations center is a system of systems that comprises the physical, media network, application, and security layers (or modules). In the following chapters, this model will be used to analyze and explain systems under discussion.

An abstraction of the flow of content from its creation to its ultimate destination, the consumer, will be presented in the next chapter.

5

The content life cycle

If all things change, then change is constant; and total chaos is complete order!

Amid the transitional chaos of the new broadcasting universe, the media life cycle remains constant. Very simply put, content must be created, assembled, distributed, and consumed. This will never change.

Any media business can be described in four phases, as shown in Figure 5.1.

To use the book publishing industry as an example, the four phases are as follows:

- Creation: Copy is written, photos taken, cover art designed.
- Assembly: These creative elements are assembled into a book.
- Distribution: The book is shipped to outlets, advertised, and promoted.
- Consumption: Someone buys and reads the book.

Technological advances play a major role in each of the processes and work-flows required in each phase. New technology may open new opportunities.

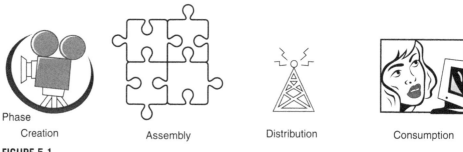

Phase
Creation   Assembly   Distribution   Consumption

FIGURE 5.1

The four-phase content life cycle

The cell phone as a means of video consumption, the Internet as a distribution channel for network TV programming, file-based workflows during the assembly phase, and images captured using modern charge coupled device (CCD) cameras have all had a positive impact.

This chapter will discuss the flow of content through each of the four phases. It is helpful to break down the broadcast chain into these phases in order to better understand the workflows, processes, and technology particular to each phase.

The broadcast TV chain will be used to establish a baseline, and then we'll look at how repurposing content for or from multiplatform, particularly the Internet and handheld devices, impact each phase in the content life cycle.

DISRUPTION OR EVOLUTION?

Journalists often use the term "disruptive" to describe the recent state of the media business. The transition to digital media has fundamentally transformed broadcast operations. Nearly all production processes are digital; only legacy content remains as analog form. Eventually, even this will all be converted to digital.

Rather than continuing to be disrupted, the media industry is adapting and is really at a pivotal moment. The three-screen universe is about opportunity. And TV broadcasters are in the pole position as the race begins. It may be fashionable to herald the demise of television, and in some ways this may be true. News consumption has definitely migrated away from television and newspapers to the Internet.

The Super Bowl will always be a mass audience event. I may watch the pregame show on my cell phone while I am on a train just to get commentary from my favorite analysts. But I, along with millions of others, will arrange my day so that I can be in front of my HDTV at kickoff time. I'm willing to wager that many will pay to see the game in a movie theater in 3D, as it was presented at an invitation-only test event in 2009. I can't imagine that anyone would choose to watch the game on a 2.5 inch LCD screen when a better alternative is available.

Television broadcast signal chain

The flow of content through the TV air chain was once relatively straightforward: remote backhaul, studio cameras, and VTRs or telecines were the sources; the program control room (PCR) and master control room (MCR) switched and assembled program elements; this "program" signal modulated a carrier and was transmitted to a receiver. This was and, on a fundamental level, still is the broadcast air chain.

As with just about all aspects of life in the new millennium, today, with the digital communications revolution, each phase of the content life cycle has broadened, expanded, and become exponentially more complex.

Figure 5.2 shows how content moves through each phase of the TV operations chain. It closely adheres to the four phases of the content life cycle.

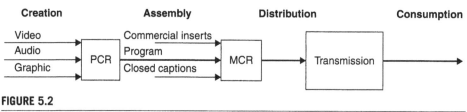

FIGURE 5.2

The content life cycle for television

THE CROSS-PLATFORM MEDIA LIFE CYCLE

With 100 HD channels available and so many complaining that there is very little worth watching, compelling content is the key to attracting viewers. More and more people are turning to TV simulcasts, reruns, and video over the Internet. Original programming is being produced for the Internet, and a few of these shows have migrated to television. Specialized content is now being produced for cell phone delivery. Sports, weather, and news alerts can automatically be sent to your PC or cell.

From one perspective, television can be considered the driver of content production for multichannel distribution. On the other hand, the low cost of entry to Internet broadcasting has enabled anyone with the guts to create content and to be able to instantly get it out to the world.

Making money from new media channels is another story. All broadcasters will admit that alternate distribution modes, regardless of all the hype, have yet to be profitable. As NBC Universal CEO Jeff Zucker put it in early 2008, "Our challenge with all these [new-media] ventures is to effectively monetize them so that we do not end up trading analog dollars for digital pennies."

"Produce once and distribute everywhere" is the mantra. This requires an agile BOC infrastructure where format conversions, compression, packetization, and modulation occur as automated processes that are "aware" of the distribution channel. Content is now regularly reformatted for the Internet, PDA, cell phones and, by a few innovative organizations, mobile DTV.

Yet with all the talk of disruption, the content life cycle still applies to all new media scenarios. Whether individually, or collectively, content is still created, assembled, distributed, and consumed. It is a powerful model to keep in mind as infrastructures evolve to support all three screens: TV, PC, and handheld.

Figure 5.3 expands the four phases to other delivery channels and delivery devices. As you can see, there are many common processes.

Once again, the big-picture view is an oversimplification. An infrastructure that supports three screens is within reach. But support of interactive television, enhanced television, gaming, product placement, and targeted, personalized features requires new technology and production techniques. These features are being deployed now in test markets.

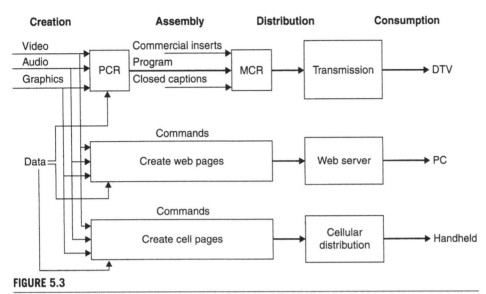

FIGURE 5.3

Multiplatform production and distribution and the four-phase content life cycle

The creative phase

Phase one is the creation of content. Storytelling is what the creative phase is all about. Each segment of a news show is a story. Sporting events are unscripted drama. Even a game show tells a story. Compelling graphics can hold viewers' eyes on the DTV set. Multichannel sound is immersive. And it can be hard to take your eyes off an HDTV image.

Audio, video, and graphic elements—"essence" in contemporary jargon—can be acquired live from a remote site or locally in a studio, produced, and stored in a file or on tape or transferred from film using a telecine machine.

A single NTSC production-based process has given way to multiple format sources. Consider a news show that uses citizen journalists providing HDV, or Internet video, as well as professional SD 480i for an HD news broadcast. Sophisticated graphics and animations as well as 5.1 surround sound are required for aesthetically compelling content. Set design and show performance have escalated to the high production values of an immersive theatrical event.

Let's consider the technology that supports each of these sources.

On the set

Television studio sets have come a long way since the days of Soupy Sales, when scenes were painted on canvas frames and props hung from the woodwork. Today's sets are more like Broadway show or rock concert multimedia productions. Watched any studio origination television lately? Then you've noticed that studio sets have become amalgams of displays, lighting, and virtual elements.

Still, no matter how perfect a digital production and transmission infrastructure might be, no matter how perfectly a consumer device reproduces the data it

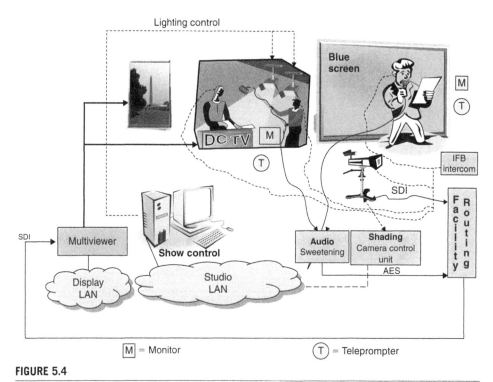

FIGURE 5.4

Functional components of a high-end set, studio, and show control infrastructure

receives, the quality of the presentation will only be a good as the source. What you see (and hear) is what you get, so you better make sure the source is of the highest production quality.

Broadcast engineers need to be aware of the technical requirements of a modern studio environment. Studio technology is now just as complex as any other system in the four-layer BOC infrastructure. The layers within a studio consist of physical components (lighting, displays, and props), media network (control communications and content movement), applications (computer platforms that run show control programs), and security (locked studio and control room doors and control system log on with passwords).

Show control, multivideo displays abound. Systems found in a studio are illustrated in Figure 5.4. Beyond the usual camera control unit, there is the addition of show control, multivideo processors, and other networked equipment installed in computer workstations.

It's all in the telling...

Creativity is no longer limited by technology. With CG, GFX, and animation applications anything is visually possible. The same is true for audio. But one must not lose sight of the storytelling mission. Do the GFXs and sound contribute to the narrative? Does the graphic convey appropriate information that is easily understood? We often hear criticism about implementing technology for technology's sake, to have the latest and the greatest. A similar argument can be made about GFX, visual, and audio effects for creativity's sake.

Studio design

HD studios require meticulous attention to visual detail when building sets and installing video displays and lighting. Today sets are dynamic, made of Plexiglas and metal, and designed in a way that the use of lighting can produce any color and effect desired. These futuristic designs and radiant colors exploit the new capabilities of DTV.

With the increased sophistication of set technology, precise control of all show elements is required. Makeup, lighting, set design, show control, and set construction (either real or virtual) are all integral parts of twenty-first century TV production and have evolved to meet the requirements of HD.

Make up

There is a story that has circulated since 1998 when the first HD broadcasts were going on the air. An anchorwoman was overheard as being totally against HDTV because she thought it would bring out every blemish on her face.

Psycho visual tests have shown that the visual perception is extremely critical of fidelity in skin tones. Proper application of makeup is so important to HDTV production that the Television Academy of Arts & Sciences has presented seminars discussing HD make up techniques.

Test scenes available for camera calibration include images of real people representing the gamut of skin tones. In addition, modern HDTV cameras have built-in skin-tone correction circuits.

So, contrary to the anchorwoman's worries, new airbrush HD makeup techniques, improved lighting, and electronic skin-tone correction actually make talent look better than they ever did before!

Lighting and show control

Lighting effects that accentuate the dramatic action have been an integral part of studio production since day one. Color media technology has progressed from simple colored gels to color temperature-balanced correction filters; lighting control has gone from manual operation to sophisticated computer automation.

A distinction must be made between overall control and the control of individual resources. Industry jargon uses two terms: "entertainment control" and "show control." Entertainment control systems operate a particular type of resource such as lighting, sound, video displays, rigging, or even pyrotechnics. Show control links together the various entertainment control systems.

Two serial communication protocols are predominately used for lighting and show control, DMX512 and MIDI show control (MSC). Connections are made over USB, musical instrument digital interface (MIDI), or LAN.

A worldwide standard, DMX512 is used by lighting consoles to send information to dimmers. Intensity levels, color changers, automated light sources, and smoke machines use the DMX512 protocol. Information is sent using 8-bit digital codes. Up to 512 devices at unique addresses can be controlled over one cable. Motorized lights require fine adjustment than 256 levels so more than one DMX address is used to extend the range to 65,536 levels.

MSC is an extension to MIDI and uses system exclusive commands. MSC can synchronize lighting cues, music playback, set element motion control, and other show control devices. Commands are accurate to within 1/30th of a second (frame accurate). Systems can be used to control lighting and sound cues to create a sequence that will be identical every time.

Field origination

A remote broadcast truck, frequently an 18-wheeler, is actually a mini-broadcast operations center up to the production control room. A completely produced program is backhauled to the operations center. There, commercials, logos, and closed captions are mixed with the clean program feed.

Electronic newsgathering (ENG) vehicles are smaller than broadcast trucks, often vans. Production capabilities are not as elaborate as a broadcast truck. However, ENG vans usually have transmission capabilities, including a microwave/satellite dish on the roof.

Audio and video formats

Two other issues to consider that are relevant to the creation phase are audio and video acquisition formats and metadata.

Digital audio and video come in more than one flavor. Format conversion is to be avoided, but invariably will occur. To enable efficient workflows, a house-standard audio and video format is decided upon. Many broadcasters have found that by converting all incoming content to the house format, conversions can be kept to a minimum, with subsequent simplification of workflows and reduced expense for conversion equipment. Figure 5.5 illustrates the method.

Unfortunately, in a digital media environment, once content has been converted into a file, the ability to read a label to identify content disappears. Content must be tagged with metadata as it is created and ingested into the asset management system. A major challenge facing the broadcast industry is to be able to trace digital content via metadata in either direction, through the four phases of the content life cycle.

With production facilities and processes being spread across the country and around the world, new methods of distributed real-time collaborative capabilities are being developed. These content delivery networks are dedicated infrastructures

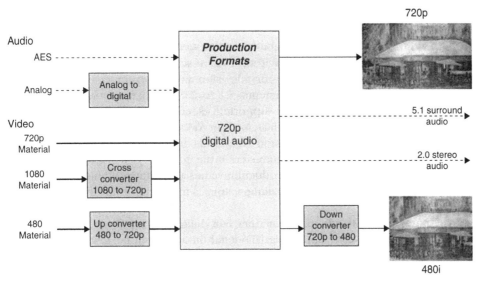

FIGURE 5.5

Conversion of all incoming audio and video to house format

systems with guaranteed data rates. Successful experiments have been delivered in uncompressed HD.

Multiplatform creative

As has just been described, producing content for television is a huge undertaking. Unless you've actually been in a studio and toured the technical operation of a facility, it is difficult to get an idea of just how large-scale it is.

To be a live Internet broadcaster, all you need is a computer, a webcam, and a microphone; anyone can do it. PC desktop software enables a person of any skill level to create their latest epic with the ease of drag-and-drop interaction and post it to YouTube. Internet radio broadcasting is even easier: just convert to MP3 and stream away; many hosting services are available for reasonable fees.

However, creating and preparing broadcast-quality content for distribution over multiple channels poses many technical challenges. Presentation issues top the list: audio and video must be converted to the proper format for the reception device. Compression issues and reception device compatibility issues add to the complexity.

For example, HD video uses pixel grids that are either 1920 × 1080 or 1280 × 720. Computer monitors do not natively support these formats: 1920 × 1200 or 1280 × 768 is as close as they get. In addition, widowing will require format conversion, and not a single cell phone display has attained the HD level of native pixel-grid resolution.

Similar limitations apply to 5.1 surround audio when it accompanies content delivered to a PC or cell phone. Sure those little earphones are convenient, but would you listen to them at home if you had the choice of connecting your MP3 player to your home audio system?

Format conversion is a core operation in the workflow that repurposes content for another platform. HD content will have to be scaled down, cell phone content scaled up, and Internet content up for television and down for cell phones. Audio presents similar conversion requirements: 5.1 to 2.0 stereo; mono or stereo to 5.1. Many flavors of conversion must be supported, especially where the sources of content are Web-friendly formats like JPEGs, MOV, or AVI files.

User-generated content has become ubiquitous on the Web. This impacts quality. A viewer on a PC is usually more interested in the content rather than the presentation quality. Less than the highest production values are sufficient. This is also true of TV news: getting to air first and breaking a story is more important than getting the perfect shot.

The moral of our story? Quality matters. For video, viewing distance can mitigate resolution issues; the eye cannot see HD detail on a cell phone, even if the phone were able to display it. But once the novelty of watching a rerun of your favorite TV show on your cell phone wears off, odds are you will DVR the next episode.

Program assembly

In the program assembly stage—the second phase of the content life cycle—audio, video, and graphics elements as well as edited segments and animations are assembled into a program and passed on to the distribution channel. In the analog era, when a program left master control for transmission, it was a composite signal ready for modulation on an RF carrier. In the digital age, it is a stream of data packets.

In the analog TV system, audio and video timing was innate; the program arrived preassembled. The receiver simply had to demodulate the signal and then present the audio and video.

This is not so straightforward in the DTV universe, where assembly of program elements is a lot more complicated. The audio and video exist separately as data packets and do not have an innate timing relationship. Audio and video compression introduces variable delays. Therefore, in addition to program content, assembly instructions and timing information must be sent in special packets in the program.

Content flow

Integration of audio, video, and graphics elements into a program is done primarily in the PCR. This step can be considered the doorway to the assembly phase. Insertion of logos, ratings, and closed captions, as well as signaling cues for downstream automated commercial insertion, happens in the MCR.

This is more or less the same signal flow as in the analog domain, right? Well, not exactly. First off, the audio and video are carried in an SDI stream. Audio channels may be embedded in the horizontal blanking areas (HANC) of the SDI signal.

But the biggest difference is that after master control the audio and video must be compressed. This encoding process will turn the raw digital audio and video into

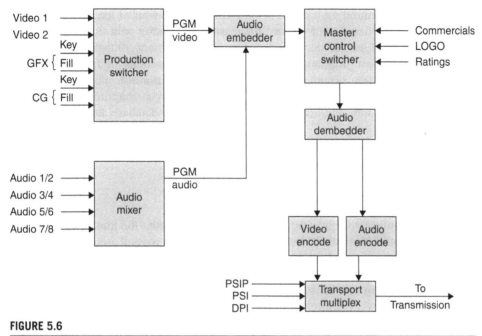

FIGURE 5.6

Assembly phase for TV operations

elementary streams (ES) and then packetized elementary streams (PES). Any implied temporal relationship between the audio and video that existed in the SDI signal is lost. Audio and video packets are marked with a presentation time stamp (PTS) for use by the DTV receiver. The PES packets are further divided into 188-byte packets and multiplexed into an MPEG-2 transport stream (TS).

Assembly instructions are multiplexed into the TS too. The technique depends on the DTV standard and the delivery channel. MPEG-based systems such as digital video broadcast (DVB) use the MPEG-2 protocol of program-specific information (PSI) while the ATSC includes MPEG tables but adds A65 program-specific information protocol (PSIP). Downstream commercial insertion is made possible by digital program insertion (DPI) splice points and cue tones that are added to MPEG-2 TS during creation of the emission multiplex.

Figure 5.6 graphically presents an overview of the assembly of a DTV program.

In the PCR and MCR, insertion and switching of audio, video, and graphics is a real-time activity, done on the physical layer, with SDI and AES being the primary essence format. Content may have been moved to playout servers over the media network in compressed format, but when taken to air, it is decompressed back to SDI and AES for switching operations.

A production switcher, loaded with macros, controls the firing of the playout servers. Command, control, and management communication occurs over a combination of LAN, RS-422, and general purpose interface (GPI). Sometimes a dedicated device is required to convert from one control format to another, say IP to GPI.

Applications running on computer platforms "fire" playout of graphics, clips, and other content sequences. For a live broadcast, the director will command the firing of a command sequence, with the rest of the segment controlled by automation. With a long-form program, with all content and interstitials on tape or servers, a redundant automation system, timed to reference signals, controls playout to air.

Some graphic elements will be assembled on the fly by a computer-based playout application. GFX, audio, and text elements will reside in a database and be assembled into templates.

The result must be seamless assembly and presentation of program elements. There are no second chances.

DTV audio and video compression

During the compression process, audio and video is coded for transmission in the appropriate codec format. In the ATSC terrestrial DTV system used in the United States, MPEG-2 is mandated by law in Part 47 of the FCC rules and regulations. Cable, DBS, and telco DTV broadcasters are free to use any codec they choose.

Perceptual compression is an enabling technology for digital transmission and production. It takes a compression ratio of over 50:1 to squeeze 1.5 Gbps HD content into a 20 Mbps MPEG-2 TS pipe. Audio is no bargain either. Raw PCM audio is about megabit per second and six channels, even with the limited frequency response of the LFE channel, and will be over 5 Mbps. The total bit rate for video and audio will be reduced to less than 512 Kbps 5.1 surround sound for transmission.

Since compression format conversion is such an important part of the content repurposing workflow, a more detailed discussion is deferred to Chapter 9.

Beside the basic issue of conversion, as mentioned earlier, variable delays in the process of compressing audio and video separately must be accounted for. Conversion systems, and PCR and MCR switchers, as well as many other pieces of audio and video equipment, have buffering capabilities that can be used to restore audio and video timing.

The MPEG-2 transport stream

The final step in the ATSC DTV assembly process is the multiplexing of audio, video, and data packets into a stream of no greater than 19.39 Mbps. The method is specified in MPEG-2 and called a Transport Stream.

The process begins with compression and concludes with the production of transport packets as shown in Figure 5.7.

MPEG-2 transport packets are 188 bytes in length and consist of two fields, as shown in Figure 5.8. The first 4 bytes are the header; the remaining 184 are the payload. Depending on the setting of particular bits in the header, the payload will be identified as consisting of audio or video or the presence of adaptation fields with additional information.

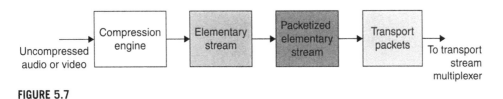

FIGURE 5.7

Processing steps to convert baseband audio or video to transport packets

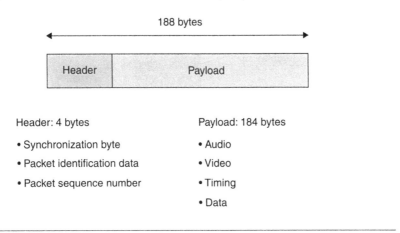

FIGURE 5.8

MPEG-2 transport stream packet structure

Key to locating and identifying the content of packets is the packet identification data, universally referred to as a PID. Packets of audio and video elements for a DTV program each have a unique PID. Each audio and video packet will have a particular PID in its header field that is used to parse the desired audio and video from the TS.

MPEG-2 TSs, as specified in ISO/IEC13818-1 is limited in the ATSC standard to data rates of 19.39 and 38.78 Mbps.

Assembly instructions

DTV assembles audio, video, and data elements for presentation at the receiver; they are transmitted without any innate temporal relationship. In fact, video frames are sent out of order.

The only means of identifying a TS packet is its PID. But this information alone is not sufficient to locate the audio, video, and system timing (clock) within the TSs that are necessary to assemble a complete program.

System timing: the system time clock

Because of the lack of an innate timing relationship between audio and video packets, a mechanism is necessary to insure that the original image and sound timing is recreated when the image and sound are presented. This is accomplished by a mechanism known as the system time clock (STC). STC packets that include a "wall clock" are part of the information in all DTV systems that use MPEG-2 TSs.

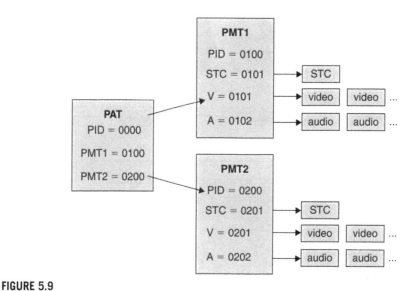

FIGURE 5.9

MPEG-2 program-specific information (PSI) tables

MPEG-2 assembly

MPEG-2 provides tools to accomplish program assembly. These are in the form of "look-up" tables found in what's referred to as PSI.

Program-specific information

PSI is a set of tables that facilitates program construction from TS packets. The program allocation table (PAT) and program map table (PMT) identify the PIDS for program elements.

As Figure 5.9 illustrates, the PAT always has a PID of 0000. This allows it to be parsed from TS without confusion or ambiguity. The PIDs of all the PMTs located in the TS are identified. There may be a single program or multiple programs and referred to as SPTS, single program TSs or MPTS, multiple program TSs, respectively.

Using the PIDs supplied in the PAT, the PMT of interest is located and then parsed to reveal the PID of the STC, video and audio packets. The decoder now simply demultiplexes the video and audio packets and sends them on to the appropriate buffer associated with decompression codec.

ATSC program assembly

MPEG PSI failed to meet the needs of American broadcasters, who needed a means to maintain long cultivated audience channel brand and number associations. They also required a method for simple navigation among multicast channels.

Program and system information protocol

One of the primary differentiators between analog television and digital television is the capability for digital systems to deliver data in addition to the audio and video.

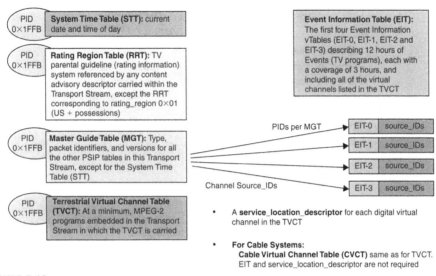

FIGURE 5.10

Required PSIP tables

This includes the mandatory carriage of Programming and System Information Protocol (PSIP) data, ATSC A/65. PSIP data includes program descriptions, ratings information, and closed captions. Presence of PSIP data is an FCC requirement.

Figure 5.10 illustrates the relationship and function of the required PSIP tables.

When the presence and content of PSIP data is carefully considered, one potential application of program descriptions found in the event information table (EIT) is that it may be used by the asset management system for classification and search. By storing the EIT and linking it with the program, the EIT can also be used when a program is rebroadcast. Similarly, RRT (ratings) data can be linked to the audio and video content and used when the program is repurposed, ideally, in an automated workflow.

If all this information were abstracted one layer above the textual implementation of EITs, RRTs, etc., and stored in a database, this database could be associated with the program content and used for repurposing in creative and unique, broadcaster-differentiating ways. This is the idea behind media objects and content wrapper techniques.

PSI and PSIP

An ATSC transmission is required to carry MEPG PSI and PSIP tables. The information must be consistent in both. Figure 5.11 illustrates the relations between methodologies.

Transport stream packet multiplex

The interleaving of audio, video, assembly, and timing packets is a form of multiplexing. Strict multiplexing sequentially alternates input data: the data is output at a rate that is the input rate multiplied by the number of data inputs.

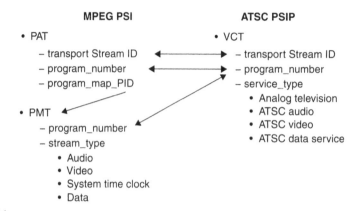

FIGURE 5.11

The relationship between MPEG PSI and ATSC PSIP methodologies

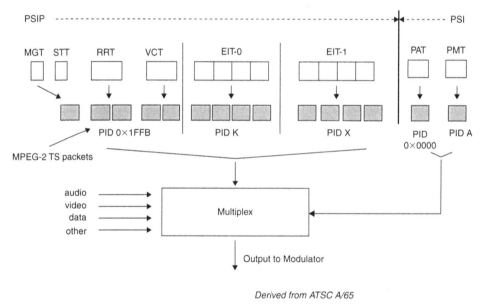

Derived from ATSC A/65

FIGURE 5.12

Audio, video, and PSIP multiplex

MPEG-2 multiplexing is not a strict multiplex. Video packets greatly outnumber audio packets; audio packets occur more frequently than PSIP or PSI packets. This underscores the importance of the PID. Packets are parsed from the stream based on their PID. Figure 5.12 is a visual representation of packets that are included in a TS multiplex.

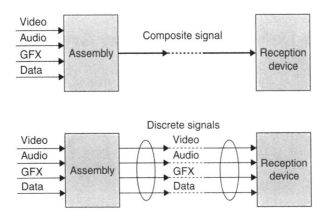

FIGURE 5.13

Assembling content for multiple delivery channels

Element assembly for multiple platforms

Assembling content elements for more than one distribution channel is complicated. Rather than brute force use of distribution channel-specific independent workflows, an elegant solution that integrates production resources and produces format ready for distribution over each channel is preferred. This approach will save capital expense on redundant equipment, it will reduce production time, and it will enable more to be done with existing staff.

The basic method strives to leverage audio, video, graphics, and data elements created in phase one. Two basic techniques can be applied to assembling content for delivery.

Compositing takes audio, video, and graphics information and produces a single, integrated indivisible stream (or file); this is similar in concept to analog television. The discrete technique delivers audio, video, and data elements independently; this is similar to an MPEG-2 multiplex.

Figure 5.13 is a functional diagram that compares the two techniques. The selection of which method is employed over a channel defines the presentation characteristics and feature set available in the reception device.

Programmatic assembly

As one might expect, with the Web being a computer- and software-based technology, software programming is the key to Web page assembly. Instructions for Web presentation are defined in the programming code used to create the page.

HTML

Hypertext markup language (HTML) is the foundation of Web page authoring. Although many other improved Web-oriented programming languages continue

The High Definition Television Archive Project

This site is dedicated to gathering the memories and impressions of those of us who designed and built the various research and prototype systems of High Definition Television as it evolved from the conceptual, through the various contending systems and culminated in the Grand Alliance ATSC standard.

The Grand Alliance Prototype HDTV ATSC System
at the Advanced Television Test Center, Alexandria, VA
Collection of Sarnoff Research Institute, Princeton, NJ
Philip J. Cianci 24"X42," Oil with Mixed Media on Wood
Copyright 2000 Jepurham Creations

FIGURE 5.14

Rich media HTML Web page with video, audio, jpeg, and text

to appear, HTML has persisted. It is simple to use to create Web pages, because there are no subtle syntax or implementation nuances. With its simplicity comes reliability.

HTML may be simple, but sophisticated multimedia Web pages can be created with it. Figure 5.14 is produced by Listing 5.1. This relative body of HTML code implements audio and video playback as well as the usual set of Web site features.

For TV broadcasts repurposed for the Web, a choice needs to be made as to whether to run the broadcast exactly as it has gone to air, or use a clean feed and mix graphics via the page code.

Cell phones offer no choice. Graphics and data will not be legible, and must be managed separately from the program stream. The only caveat is that screen resolutions for handheld devices have reached pixel-grid dimensions that approach video-cassette levels of presentation quality.

The point of this is that if audio, video, graphics, and data are kept independent of each other, the assembly phase can arrange for transmission of each element in a way best for the delivery channel.

Compression requirements for each distribution channel and target device vary. Meeting the optimal data rate raises quality issues that must be addressed. Frame rate influences data rate.

For example, MPEG-2 HD video is compressed to about 12 Mbps for over-the-air delivery. Few ISPs can guarantee sustained delivery of this data rate. Therefore, MPEG-2 video will not make over the Internet. Even an AVC or VC-1 codec, which reduces the bit rate for HD to 6 Mbps, will not enable HD over the Internet.

```
<TITLE>Iopherion Creations, Philip J. Cianci</TITLE>
<META NAME="keywords" CONTENT=
"HDTV, ATSC, Grand Alliance, Television,
Digital, Archive, Smithsonian, Cianci">
<META NAME="description" CONTENT=
"An insiders view from the benches in the trenches of
the evolution of HDTV and beyond...">
<H1><Font SIZE="12">The High Definition Television Archive
Project
</FONT></H1>
<IMG SRC="GA@ATTC.gif"ALIGN=right>
<embed src="DSCN0625.mov" width="320" height="280"
autoplay="true">
<p><BR>This site is dedicated to gathering the memories and
impressions of those of us who designed and built the various
research and prototype systems of High Definition Television as
it evolved from the conceptual, through the various contending
systems and culminated in the Grand Alliance ATSC standard.
</p>
<clear>
<p><BR><Font SIZE="4"><B><i>The Grand Alliance Prototype HDTV
ATSC System</B></FONT>
<BR>at the Advanced Television Test Center, Alexandria, VA
<BR>Collection of Sarnoff Research Institute, Princeton, NJ
<BR>Philip J. Cianci 24"X42" Oil with Mixed Media on Wood
<BR>Copyright 2000 Ioperhian Creations<o:p></i></p>
```

LISTING 5.1

Source HTML for Web page in Figure 5.14

Assembling content for cell phones and handheld devices

Programming languages have been developed specifically to ease the use of audio and video on the Internet and for delivery of content to cell phones. Synchronized multimedia integration language (SMIL) is one such language.

SMIL

SMIL is a W3C-recommended XML markup language for describing multimedia presentations. It facilitates the presentation of text, images, video, and audio. A SMIL presentation can contain links to other SMIL presentations. Markups are defined for timing, layout, animations, visual transitions, and media embedding.

SMIL files are text files, similar to HTML. Developers have a variety of tools they can use to create SMIL: dedicated authoring software and SDKs, XML editors, or text editors.

The Open Mobile Alliance (OMA) multimedia messaging service (MMS) specification defines the use of MMS SMIL for handheld applications. The benefit of using this SMIL profile is that all OMA-conforming MMS devices will be able to understand and play a presentation.

The 3GPP suite of specifications defines the 3GPP SMIL profile. This SMIL profile is designed primarily for streaming purposes (and is defined in the 3GPP Packet-Switched Streaming Services version 5 specifications, also referred to as the PSS5 SMIL profile). The 3GPP specification also permits use in MMS. Although 3GPP SMIL does not include all the features of SMIL 2.0, it allows much richer content to be created.

Distribution

HD/DTV distribution began over the air and now includes cable, satellite, and telco delivery to the home. Legal and business issues regarding retransmission consent come into play defining how terrestrial broadcasts can be distributed over these other technologies.

The Web has been a Wild West for TV programming. Broadcast content often finds its way to video sharing and social Web sites. Even the most rudimentary consumer media management and editing software includes the ability to effortlessly upload content to a Web site. But before we talk about the newest content distribution channels, let's look at the development of communications distribution from the telephone on forward.

Content distribution

When telephone was invented in 1876, telegraph lines provided a preexisting infrastructure. As a result, telephone systems were immediately operational: there was no need to build a new distribution channel from scratch. This scenario would be repeated with the consumer adoption of the World Wide Web in 1995.

Although primitive by today's standards, telephone technology was the first to use communication principles we still use today. The conversion of sound waves in air to electrical signals by a transducer (microphone) established the fundamental technique of capturing the world of our senses in an analogous manner to electricity.

Analog vs. digital modulation

Analog modulation varies the amplitude, frequency, or phase of a high-frequency sine wave. Advanced techniques can use combinations of these basic forms of modulation. The term "analog" refers to the information that is to be conveyed. Analog signals produce variations in a step-less continuum of voltages.

Digital modulation applies discrete voltage levels to the RF carrier, creating characteristic patterns of steps or "constellations" when viewed on appropriate monitoring equipment. In actuality, modulation is always an analog operation, even when the information is in the form of digital symbols.

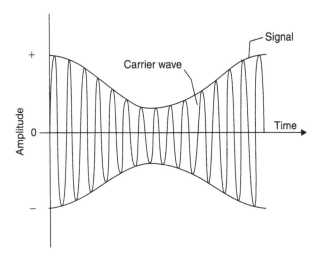

FIGURE 5.15

Amplitude modulation of a carrier wave

Amplitude modulation

Just after the turn of the twentieth century, Reginald Aubrey Fessenden invented a high-frequency alternator that produced a continuous radio wave. It was a significant improvement of the intermittent spark-gap generators that had been previously used for wireless transmissions. Fessenden invented a way to modulate the amplitude of radio waves, the enabling technology for AM radio (see Figure 5.15).

Amplitude modulation is transmitted as a sky wave: it reflects off the earth's ionosphere. The ionosphere is a layer of charged particles, theorized in 1902 by Arthur Kennelly and Oliver Heaviside, and discovered in 1924 by Edward Appleton. This reflection enables AM radio signals to travel great distances. However, changes in the ionosphere at night alter its reflecting characteristics, causing variations in signal propagation distances. Lightning is a source of electromagnetic radiation and degrades an amplitude modulated radio signal.

The invention of the vacuum tube by De Forest improved the system. But it wasn't until Armstrong developed heterodyning that radio transmission became practical.

Carrier radio waves are very high frequencies when compared to audio. This can create difficulties in accurate transmission of information if modulation is directly applied. Heterodyning is a technique that first modulates an intermediate frequency (IF) with the signal and then modulates the carrier with the IF containing the audio in the second stage. The technique has been extended to enable use of RF carriers in the gigahertz range for satellite transmission.

With the development of radio broadcasting technology, communications were no longer limited to areas where a physical cable reached. Radio waves propagated freely in space. Broadcasting was quickly monetized. Commercials paid for sponsored programming.

Television makes the scene

With the problem of capturing and broadcasting sound solved, it was only natural for researchers to turn to the sense of sight. By that time, radio broadcasting had become big business. Big corporations had money to spend on speculative R&D projects. AT&T and RCA backed rival inventors Vladimir Zworykin and Philo T. Farnsworth, respectively. When the smoke cleared, the NTSC television standard was established. Parallel efforts in Great Britain by John Logie Baird produced a television system that was on the air prior to WWII.

Although some of the initial systems were electromechanical, NTSC I, the original black and white system in the United States, used all electronic technology.

Analog TV broadcast technology

Compared to radio broadcasting, television was infinitely more complex. Radio consisted of one, relatively low-frequency signal, the human voice is below 3 kHz, while a TV signal had video at 4 MHz rates and the requirement to mix audio into the signal as well.

The introduction of color made matters all the more challenging. Color was conveyed as the difference in phase of the color subcarrier, a 3.58 MHz sine wave.

Different systems, each with variations and improvements, were developed around the world: PAL in Europe, except for France (naturally), which developed SECAM. The systems engineered by these pioneers served the world well for more than 50 years.

Digital transmission

Digital transmission systems consist of two steps. In the first, frequently called error correction and concealment (ECC), the digital data is processed to insure maximum robustness for the intended delivery channel. In the second step, ironically, the digital signal is mapped to an analog voltage and then modulates a radio frequency carrier wave.

Error correction and concealment

Distribution of real-time digital media over RF- and cable-based systems is subject to electromagnetic interference. If a transmission error occurs, lost data cannot be retransmitted. Therefore, a means must be employed to insure robust delivery and reception.

Figure 5.16 illustrates the components that are involved in the ECC step.

As shown, ECC consists of four steps. Each is tuned for maximum resiliency of the intended delivery channel technology.

Channel coding The purpose of channel coding is to make maximum use of channel bandwidth and minimize burst noise errors by spreading data over full channel. This prevents long runs of zeroes or ones.

In the ATSC DTV transmission system, the randomizer generator polynomial is:

$$G(16) = X^{16} + X^{13} + X^{12} + X^{11} + X^{7} + X^{6} + X^{3} + X + 1$$

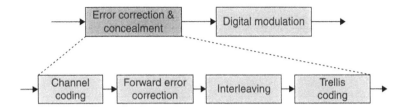

- Randomizing: channel bandwidth energy dispersal
- Forward error correction: enables repair or erroneous bits (adds extra bits)
- Interleaving: time dispersal of data
- Trellis coding: state machine-generated forward error correction (adds extra bits)

FIGURE 5.16

Distribution phase for TV forward error correction

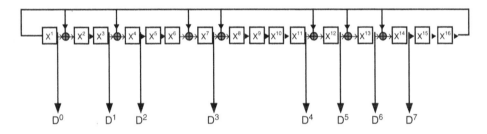

The generator is shifted with the Byte Clock and one 8-bit byte of data is extracted per cycle

FIGURE 5.17

Channel coding randomizer polynomial (using D flip flop and exclusive OR gates) (Derived from ATSC A/53)

The initialization (preload) to $0 \times F180$.

Implementation of the algorithm can be accomplished by using a shift register feedback circuit as shown in Figure 5.17.

A shift register has an input, data registers, and an output. The data register in a hardware implementation is similar to latching D flip flop (see, Appendix, Figure B.3). A bit enters the shift register, and as it traverses the chain, is exclusive ORed (XOR, see Appendix B) with the output bit as defined by the generator polynomial terms.

The result is data that has the spectral characteristics and data distribution of random noise. Since electromagnetic noise is usually limited in frequency, this technique reduces the probability that a noise burst will destroy an unrecoverable number of data words because data is distributed uniformly over the channel bandwidth.

Forward error correction Television is real time; if data is corrupted during transmission, it cannot be resent. Forward error correction (FEC) is a means of enabling detection

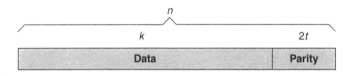

FIGURE 5.18

Data payload with Reed-Solomon FEC parity

and correction of bad data by including parity information in the transmission. DTV system use Reed-Solomon techniques.

The Reed-Solomon encoder takes a block of digital data and adds extra "redundant" bits. Errors occur during transmission or storage for a number of reasons (e.g., noise or interference, scratches on a CD, etc.). The Reed-Solomon decoder processes each block and attempts to correct errors and recover the original data. The number and type of errors that can be corrected depends on the characteristics of the Reed-Solomon code.

Reed-Solomon codes are a subset of binary-coded hexadecimal (BCH) codes and are linear block codes. A Reed-Solomon code is specified as RS(n,k) with s-bit symbols.

This means that the encoder takes k data symbols of s bits each and adds parity symbols to make an n symbol code word. There are $n - k$ parity symbols of s bits each. A Reed-Solomon decoder can correct up to t symbols that contain errors in a code word, where $2t = n - k$.

The diagram in Figure 5.18 shows a typical Reed-Solomon code word (this is known as a systematic code because the data is left unchanged and the parity symbols are appended):

Example: The Reed-Solomon code in the ATSC system is RS(207, 187) with 8-bit symbols. Each code word contains 207 code word bytes, of which 187 bytes are data and 20 bytes are parity. For this code:

$$n = 207, \quad k = 187, \quad s = 8$$

$$2t = 20, \quad t = 10$$

Therefore, an ATSC decoder can correct any 10 symbol errors in the code word: i.e., errors in up to 10 bytes anywhere in the code word can be automatically corrected. In Europe, the DVB DTV system uses RS (203, 187), $2t = 16$ resulting in the ability to correct up to 8 byte errors.

Interleaving Added protection against burst noise is accomplished by spreading the bits in each byte over time. The technique is known as interleaving.

Consider a sequence of bytes, A, B, C,..., where each bit is denoted by A0, A1, A2, ... The transmission sequence would be:

A0,A1,A2,...A7, B0, B1, B2,...B7, C0, C1, C2...

A noise burst of 16 bits in length would completely destroy either two bytes or one completely and two partially. Depending on the number of bits per byte that can be corrected, this damage may be beyond the ability of the FEC algorithm to correct.

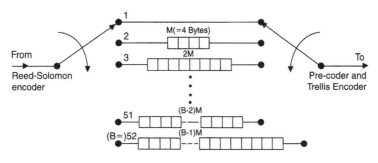

M = 4, B = 52, N = 208, RS Block = 207, BXM = N

FIGURE 5.19

ATSC Convolutional interleaver (Derived from ATSC A/53)

Interleaving the same sequence produces the bitstream A0, B0, C0, ... A1, B1, C1, ... A2, B2, B2, ...

If the same 16-bit noise burst damages the interleaved stream, the maximum number of bits damaged in any byte would be limited to two. By spreading the bits out farther, eventually a point would be reached where damaged bytes would be spread such that the damage would be spread over a number of transport packets. This helps to insure that the number of bytes corrupted by channel noise will be within the capability of the FEC implementation.

Interleaving is accomplished by a convolutional technique. Figure 5.19 shows the convolutional interleaver used in the ATCS DTV transmission system.

Bits enter shift registers 1 through 52 in sequential order, moving to the next register when the previous one is full. At the output, bits are read out in a rotating order, always from 1 to 52. This creates a completely deterministic bit pattern. At the receiver the process is reversed.

Trellis coding The final ECC processing step, trellis coding, increases the probability of a symbols occurrence. The technique also maps symbols to analog voltage levels. Figure 5.20 shows the processing blocks of the trellis coding algorithm used by ATSC 8VSB.

Trellis coding is specified by the number of input bits to output bits. In the ATSC technique this is denoted by (2,3); for every two input bits an output symbol of three bits is created.

The result is a reduction of legal locations a bit may occupy. Figure 5.21 shows the method as applied to quadrature amplitude modulation (QAM).

As the figure shows, in the topmost constellation, there are four-bit locations in each quadrant. After trellis coding, as can be plainly seen in the bottom row, only one-bit position is valid in each quadrant. Therefore, the area where a valid bit may be located has been increased considerably.

Greater than the sum of its parts

ECC is a careful balancing act of processing modules and parameters. Its effectiveness can be measured as a "coding gain." This is the difference in signal-to-noise ratio

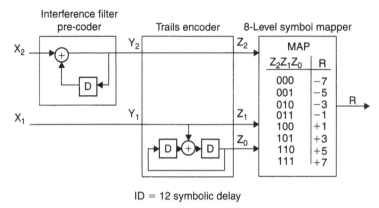

ID = 12 symbolic delay

FIGURE 5.20

Main service trellis encoder, precoder, and symbol mapper (Derived from ATSC A/53)

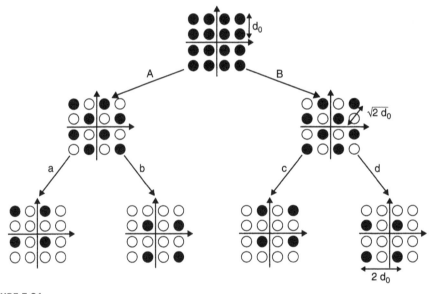

FIGURE 5.21

Modification of 16-QAM constellation by trellis coding to reduce valid data one point to one per quadrant, thereby improving transmission error resilience

(S/N) between a coded and uncoded system of the same information rate that produces the same error probability.

Digital modulation techniques

Different modulation techniques are used over each TV delivery channel. They are vestigial sideband (VSB), QAM, and quaternary phase-shift keying (QPSK).

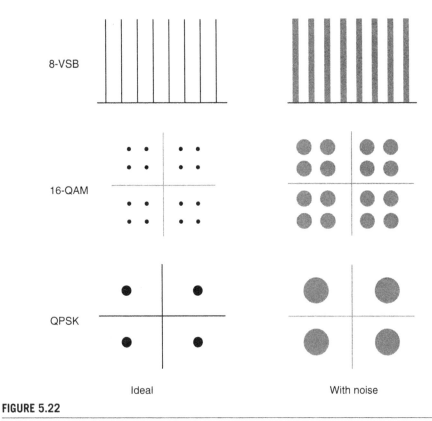

8-VSB

16-QAM

QPSK

Ideal   With noise

FIGURE 5.22

Digital modulation: VSB, QAM, and QPSK constellations and the effect of random noise

Figure 5.22 compares the different constellations produced by each method. On the right side of the illustration, it is clear that even if the signal is corrupted by noise, because the information is digital, it assumes discrete locations and innately offers some degree of noise immunity. An analog TV signal would start to look "snowy."

There is an irony to the term "digital modulation." In actuality, digital symbols of a defined bit length are mapped to voltage levels and modulate a carrier wave, exactly the same as in analog modulation. The difference is that analog modulation can be any voltage in the signal range; digital modulation is restricted to a limited number of voltage levels.

8-Vestigial sideband modulation

The ATSC DTV transmission system uses 8-level VSB or 8VSB modulation. The eight indicates that there are eight distinct voltage levels corresponding to symbols. Symbols consist of 3 bits of data.

Figure 5.23 illustrates how the eight levels modulate the carrier wave. Notice how both the top and bottom of the carrier way reflect the value of the symbol.

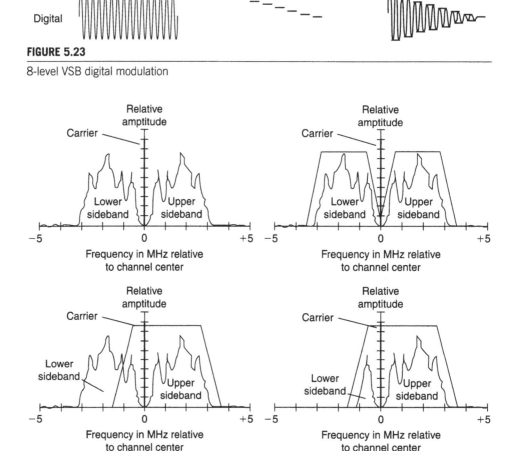

FIGURE 5.23

8-level VSB digital modulation

FIGURE 5.24

Vestigial sideband filtering

In amplitude modulation, this will create additional signals. These sidebands expand the range of the signal about the carrier wave as shown in Figure 5.24.

The use of the term "vestigial" refers to the lower sideband that is mostly filtered out before transmission. A small vestige remains, as shown in the lower right.

Quadrature amplitude modulation

QAM is a technique where quadrature carriers, waves with a phase difference of 90°, are amplitude modulated. The modulation scheme is used extensively in cable systems.

Figure 5.25 shows how the two waves are amplitude modulated and combined and then produce a constellation.

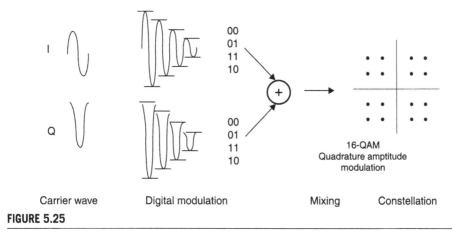

Carrier wave   Digital modulation   Mixing   Constellation

FIGURE 5.25

Digital 16-QAM modulation encodes two quadrature (a phase difference of 90 degrees) with two bit binary symbols 00, 01, 11, and 01

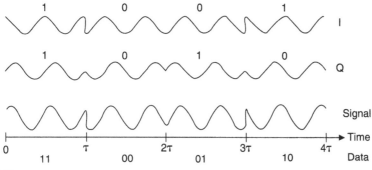

FIGURE 5.26

Quaternary phase-shift keying

The constellation is characteristic of the number of voltage levels that modulate each carrier. This example shows 16-QAM and has 4-bit points in each quadrant. A 64-QAM constellation would have 16-bit points in each quadrant. Today, 256-QAM modulation is commonly utilized.

Quaternary phase-shift keying

Satellite systems use QPSK. Figure 5.26 shows how 2-bit symbols are mapped to the phase of the carrier signal.

The resultant constellation, shown in Figure 5.27, has a data point in each quadrant. QPSK can also use more than four phases, which results in a higher data-carrying capability.

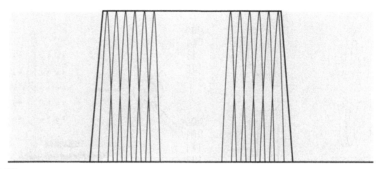

FIGURE 5.27

Coded orthogonal frequency division multiplexing divides a channel into many carrier waves, a spread spectrum technique, and modulates each carrier with a symbol

COFDM

The modulation method implemented in DVB-T systems is coded orthogonal frequency division multiplex (COFDM). A delivery channel is populated with many narrow bandwidth carriers as shown in Figure 5.27.

Each carrier contains a single-coded symbol of information. The improvement in noise immunity occurs because the symbol data is valid for a longer period of time than in other digital modulation techniques. This minimizes the potential for a bit to be totally corrupted by a noise burst, based on the concept that the noise burst will be significantly shorter than the time length that data is valid.

Multiple platform distribution

As we have been discussing, TV content distribution has expanded to include every conceivable delivery platform. Television, regardless of the compression engine, uses an MPEG-2 TS to feed the distribution technology. This standardization simplifies assembly.

Data rate must be adjusted for the delivery channel and consumption device. A DTV MPEG-2 TS is nearly 20 Mbps; this does not fit Internet distribution capacity.

MPEG-2 TS packet IP encapsulation

One approach that is used when data rates permit is to encapsulate MPEG-2 transport packets in an IP packet. Figure 5.28 shows how seven DTV transport packets are packed into an Ethernet frame.

Of course, the caveat is that network bandwidth must be able to provide an adequate, sustainable data rate. Over GigE, this is not really an issue for a 20 Mbps HDTV payload, but for the public Internet with data rate below 10 Mbps, this is impossible.

This technique can be used for Internet and handheld TV content delivery. It is becoming more feasible as advance compression codecs produce lower-bit rate, higher-quality content streams. Although HD resolution is in the 8 Mbps area, SD programs can be compressed to 1 Mbps and audio can be delivered via MP3.

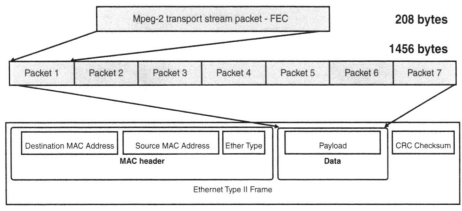

FIGURE 5.28

MPEG-2 transport packet Ethernet encapsulation

Internet bandwidth will continue to increase, while compression codecs will become even more efficient.

Network signaling

Internet broadcasting and streaming to handheld devices use real-time and group networking techniques. The relevant protocols are:

- RTSP: Real-Time Streaming Protocol
- RTP: Real Time Protocol
- IGMP: Internet Group Management Protocol

Figure 5.29 presents the protocols used to deliver video over the Internet.

In contrast to DTV delivery, Internet media delivery does not use any form of FEC. In fact, even TCP retransmission control has been abandoned for UDP, in an effort to squeeze every bit of delivery bandwidth.

Content consumption

Consumer enjoyment of the original work of art, message, or signal—that is, consumption—is our *raison d'être*. TV content is everywhere. VOD, PPV, and recording on a DVR enable consumers to enjoy their TV experience whenever they choose. This has transformed media consumption from by appointment to on demand.

Figure 5.30 shows a typical TV scenario from just a short time in the past. Each set was an island unto itself. A DVD or VCD was necessary either to view programs at a time other than when they were broadcast or to transport content from one television to another.

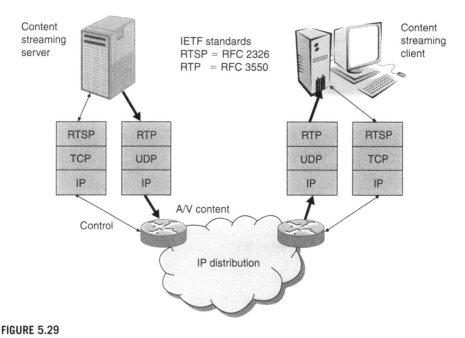

FIGURE 5.29

TV content distribution over the Internet

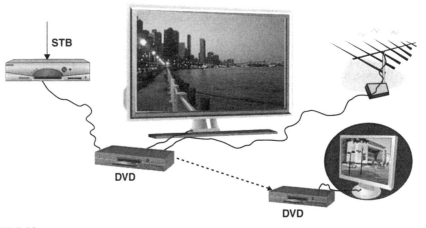

FIGURE 5.30

A multiple-room TV-viewing environment

Distribution of TV programming to multiple televisions in various rooms is rapidly being enabled. The limitations of a single set-top box and a single DVR are disappearing as service providers install the latest generation of multiroom, multirecord/playback STBs.

FIGURE 5.31

The interconnected media consumer

Consuming content on multiple devices

Personal home digital networks are transforming media consumption. Networked home media has accustomed the early-adopting consumer to being pampered and getting what they want, where they want it, and when they want it. Transparent transcoding of content format in the home network is being facilitated with increasing ease. And of course, "I" should be able to exercise my "fair use" rights to copy content. But creators' rights must be protected across platforms in the home network.

Figure 5.31 shows just a sample of how the consumption environment has expanded. The long-ago predicted networked media home is at hand.

Content transfer among devices

A key issue to resolve before a truly interconnected home media network proves to be user friendly is the need to establish device interconnection methods. This includes the physical layer and the protocol stack that rides over it. In an effort to remove this bottleneck, a number of technologies are under development and vying for a position in the connected consumer's home.

Ease of installation concerns have led to development of techniques that use the existing wired infrastructure. Every house has electricity available at outlets in every room. HomePlug and other powerline-based technologies can distribute content over the existing power infrastructure.

Similarly, nearly every home has been wired for landline telephones. The Home Phoneline Network Alliance (HomePNA) uses existing twisted pair residential phonelines for data transmission. Data rates of 200 Mbps are realistic.

With the large installed base of cable TV subscribers, use of existing coax is another option. The Multimedia over Coax Alliance (MoCA) consists of major MSOs and is a viable solution for in-residence content distribution even if it is delivered to the home over a fiber optic network.

In an ideal home networking environment, no cables would be necessary. Wi-Fi content distribution is possible depending on the required data rate. Consumer solutions are just beginning to appear on the market.

There's a lot more about consumer connectivity in Book 2.

RISING TO THE CHALLENGE

The challenge for those in the production and transmission domains is to make the creation/assembly/distribution workflow efficient for consumption on any consumer device. This takes coordinated business, creative, and technological planning.

Workflow analysis, planning, and teamwork

6

These are exciting, challenging times in the media industry. We are dealing with evolving, often unproven technologies while standards are emerging. Developing multidisciplinary teams for design and deployment of infrastructures that integrate broadcast and IT systems has placed added emphasis on project management and teamwork.

A BOC infrastructure integrates two major technologies of the twentieth century: electrical engineering and computer science. Just as the debate 20 years ago over whether computer science is an engineering discipline has been laid to rest, so too has the broadcast vs. IT debate been resolved in the emergence of a new engineering discipline that combines elements of the two: media system engineering.

Prior to networked, IT-based production, broadcast systems were divided by functionality and stood relatively alone. Now, they connect with each other over an IT infrastructure, and each system increasingly interacts with other systems. If the infrastructure, a system of systems, is planned intelligently, with an eye toward new media, it can be designed and implemented as an adaptable infrastructure that can meet unforeseen future needs.

A THREE-DIMENSIONAL PERSPECTIVE

Broadcasters have had to keep their eyes wide open on all fronts. The universe around them is disintegrating to give rise to a new balance among the pillars of the media industry's foundation.

Remaining agile enough to keep up with rapidly emerging distribution platforms in a business that demands nearly 100% uptime is at odds with existing broadcast engineering philosophy. Usually, a technology is tested until all are completely

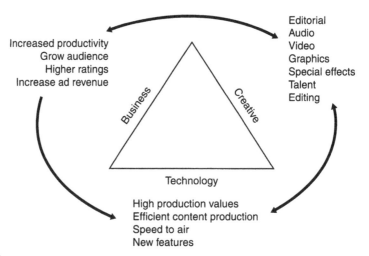

FIGURE 6.1

The symbiotic relationship among creative, technology, and business dimensions. Managing the relationships among these domains can lead to an agile, adaptive media organization

confident in its operation. Many of the techniques and technologies in use have been around for decades.

But with new media systems often being put together in short periods of time and commodity equipment placed on the air with minimal testing, broadcast engineering is unnerved. Yet the opportunities appear so quickly that there just isn't time to do 100% testing.

Better planning and improved teamwork can help resolve this issue. The key: work smarter.

The media industry can be viewed from three perspectives: business, creative, and technology (Figure 6.1).

Technology integration

The technical challenge of building an integrated multiplatform production infrastructure that supports diversified distribution technologies delivered to a wide range of consumer platforms with varying functionality is unprecedented; transitioning to DTV transmission has been relatively straightforward by comparison.

Requirements for a multiplatform infrastructure

A broadcast infrastructure is a system of systems. Some of the systems that must be integrated include the following:

- Content creation and acquisition: audio, video, backhaul, ENG, live events;
- Production: audio and video editing, graphics;
- Content management: storage and archive applications;

- Production control: audio, video, and graphics for live events;
- Master control: automation; inserting national and local commercials, promos, logos, ratings information, and closed captioning.

Once content is assembled, the workflow diverges. Formatting and assembling content for each platform includes three processes:

- Transcoding: converting audio and video to distribution formats;
- Attaching presentation assembly instructions for each channel;
- Channel delivery: converting to the required format (e.g. MPEG-2, AVC, VC-1)

The content is then transferred over the appropriate distribution channel to the targeted receiver, a DTV, PC, or handheld device.

Creative opportunities

Artistic creativity, like it or not, is directly dependent on technology. From the first cave artists to digital graphics designers, technology has been the key to new modes of expression. Witness how Gutenberg's printing press enabled the mass production of books. Oil paint in tubes gave rise to *en plein aire* painting and the impressionists. Edison's contributions of audio and video recording gave birth to the gramophone and motion pictures. The ability to tape TV programming on a VCR changed the media consumers' world, and put them in charge of when and where they watched TV programs.

Probably, the biggest boon to contemporary creativity has been the development of computer technology. Effects that took weeks to do in analog mediums now take hours in the digital domain, radically speeding up the time it takes to get such content into the production and distribution pipeline.

Artists are dependent on technology, and they know how production works. They need to work with technical staff, who generally haven't got a clue about the creative process, to come up with ideas and approaches that best serve multiplatform production needs.

Business models

The methods by which broadcasters make money are fairly well established. Terrestrial TV and radio broadcasts are free; operations are ad supported. Cable and satellite use a dual revenue model, charging the consumer for basic service and tiers of premium channels in addition to charging advertisers for presentation of commercials.

Internet service providers charge for access to the Web. These are the Telcos and MSOs of the world. Once on the Web, many services are free and ad supported (by annoying pop-ups or pre-rolls); other Web sites and services like iTunes charge for their content and services.

Cell phone fees are an extension of traditional monthly telephone charges, but they are now based on usage minutes. The key difference is that under the old landline model, only the caller was billed for the call; now both the caller and the recipient incur charges. Basic service is supplemented by text messaging, personalized ring tones, Internet access, and the introduction of new features. Consumers pay to use all of these services either by the minute or in monthly data packages.

Naturally, competition influences the price of any of these services, at least in a free-market environment. Business models must be flexible, built on a solid long-term foundation, but able to adapt to new revenue opportunities generated by new technologies, new distribution channels, and location consumption preferences. Forward-looking media companies and content providers see these as new revenue sources.

Is the business model evolving from free or subscription-based over-the-air cable and DBS to an a la carte, tiered, on-demand, or pay per view value proposition? Where does local broadcasting fit in? Is there a role and business model for SD multicasting? Where does government regulation interfere or aid? These are some of the questions that will be answered in time, only to be replaced with a new set.

For the moment, at least, the consensus seems to be that the primary business model of the future will be an ad-supported one, though it likely will not look like the traditional TV model. Time-shifting, ad-skipping DVRs threaten broadcasting's fundamental revenue stream; the 30-second advertising spot will need to adapt to emerging new consumption habits and platforms. The enlightened see this as an opportunity and are working on new ways to get product and branding messages to consumers.

With so many media outlets to choose from, it is in the best interest of broadcasters to devote attention to new cross-platform promotional techniques that steer viewers to their portfolio of media properties. When consumers are "left to their own devices," they probably will not make an effort to use anything other than a device they are familiar with, so giving them familiar content on new devices and adding supplemental new content is a way to encourage them to make the platform migration.

Program and commercial content can be personalized and targeted; that is presented based on viewer profiles, consumption history, or demographic data, or with interactive, viewer-selected features. Personalization and targeted ads make management of the viewer database a critical core business process. This viewer information must be analyzed and mined, then integrated with automated content playout and commercial insertion systems.

Advertising-based revenue models have been platform targeted. Sellers and buyers concentrated on one medium, and so, ad sales have been medium segregated. Some forward-looking companies are now offering a mixed media package; the ad package is a combination of television, radio, print, and billboards. A database-driven, 360-degree customer view is developing.

The challenges presented in each of the three dimensions are actually opportunities. They present the experienced professional the chance to revitalize his or her vocation with a child's thirst for knowledge. Think of this as you thought of your career when you began it. Be excited that you get to enjoy the discovery of new knowledge. One key to eternal youth is the perception of the world as a wondrous place with endless secrets to reveal. The evolution to the new media universe offers this stimulation daily.

According to the Merriam-Webster Online Dictionary, engineering is "(a) the application of science and mathematics by which the properties of matter and the sources of energy in nature are made useful to people, (b) the design and manufacture of complex products (software engineering)." We might add "within financial constraints, on schedule, and to fulfill a business need." The following sections will look at how engineering models and methods can be applied to increase both efficiency and productivity.

WORKFLOW DOCUMENTATION AND PROCESS MAPPING

With the increased complexity of technology, it is important to be sure that the infrastructure is efficiently designed for the production workflow. Mapping of production workflows with respect to the underlying infrastructure is necessary to engineer an optimized system design and deployment. This will result in optimum production efficiency and increased ROI in infrastructure resources.

As with any complicated undertaking, analysis and planning are necessary to ensure the desired outcome is attained. Process efficiency reduces organizational stress; everybody gets more done while seemingly exerting less energy.

The job descriptions of all broadcasting and media business personnel are expanding as new technologies are introduced, and production needs and distribution channels increase. This makes it crucial to match workflow with the supporting technology infrastructure.

A mapping of the workflow steps and processes involved is the starting point. Workflow modeling techniques create process maps.

Workflow process mapping is the documentation of steps in any task. These can be based on observation for an existing workflow or developed from scratch for a desired workflow. The ultimate goal is to identify areas where a process can be improved and make the entire workflow more efficient and productive.

Figure 6.2 is a simple example that documents the workflow of a request for a new computer-generated animation sequence.

Someone in production or remote production has the idea for a new computer-generated animation sequence. The concept is passed on to creative services, where a conceptual storyboard is fleshed out. The graphics department is given the conceptual storyboard to develop. The resultant storyboard goes through the review process. If approved, the storyboard is passed back, with changes, to the graphics department for production. If it is not approved, the process restarts at the beginning.

This map documents a workflow and not the underlying technology. The map can be used to identify and eliminate bottlenecks, and evaluate the potential improvements in efficiency that might be achieved by applying technological solutions.

Figure 6.3 is a process map of the graphics workflow from conception to air, including the technology involved in each step.

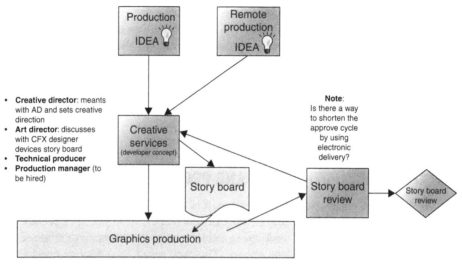

FIGURE 6.2

A process map of the first steps in a computer-generated animation-creation workflow

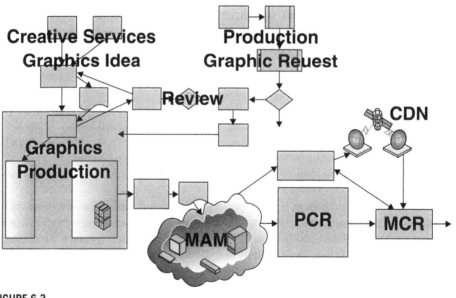

FIGURE 6.3

The complete graphics process, from conception to air

The goal of workflow and process mapping is to improve efficiency and reduce errors. In the design phases, it is meant to ensure that the underlying infrastructure performs the necessary functions to meet the project's requirements.

Bottom-up Process Improvement

IBM, in its heyday, was famous for rewarding its employees for contributions that improved processes. A side effect: employees have genuine interest in and take pride in their work.

WORKFLOW AND PROCESS ABSTRACTION

A natural animosity often exists between various departments within an organization. Each profession has built a belief system that fits its worldview and career view. Simplistically (and perhaps stereotypically), business is all about money, technology is about precision and problem solving, and creativity is about freedom and inspiration.

In the media business, a successful organization will evolve a culture where these three groups, driven by divergent motivations, all work together. If we phrase it in the form of an equation in terms of workflow or process modeling, it looks like this:

$$\text{Media Systems Process Modeling} = \text{BPM (Business Process Modeling)} \\ + \text{CPM (Creative Process Modeling)} \\ + \text{TPM (Technology Process Modeling)}$$

Modeling and abstraction are fundamental methods used by the scientific community to explain natural phenomena, frequently expressed in mathematical forms.

Take, for example, Galileo's equation for the distance traversed by an object moving at a constant speed. The equation "distance equals velocity multiplied by time ($d = vt$)" accurately describes, or models, simple motion. Under different circumstances, such as at velocities that are very close to the speed of light, this model breaks down and another takes its place. The Lorentz transformation mathematically models and explains motion at relativistic velocities.

Each model is verified by observation.

Workflow and process abstraction is an analysis method that can lead to efficiently aligning an organization with its mission.

An enterprise approach promotes business effectiveness and efficiency while striving for innovation, flexibility, and integration with technology. This is the state of the media industry today.

What follows are examples of models that can be applied to integrated media systems environments in order to maximize efficiency and productivity.

CMMI and engineering process maturity

One way to gauge organizational technological proficiency is to use Capability Maturity Model Integration (CMMI). This organizational technology competency analysis tool was developed by the Software Engineering Institute at Carnegie Mellon University in the mid-1980s. It was used by the military when evaluating the probability of success on a software project by responders to RFPs.

The technique defines five levels of organizational competence (capability) ranging from the "Wild West" to progressive and forward looking.

Level 1, the "Initial" level, is characterized by an ad hoc approach to tasks or projects. Many broadcast engineering and support departments fall into this category because of the necessity to stay on the air at all costs. A single engineer, or a very small group, generally does system design and commissioning; therefore, formal processes do not exist and information tends to be siloed.

At Level 2, "Repeatable" processes are followed throughout the life of a project. Some form of formal project management is installed. This often includes developing scheduling and budget management tools. Projects are planned and an attempt is made to adhere to the development process. This may include brainstorming sessions, design reviews and improved documentation, and the communication of milestone accomplishments to management.

Level 3 is characterized by following "Defined" processes for all projects. Processes are developed and applied on an organizational scope, not just within a single department. A project management office (PMO) may have been established to develop and manage standardized processes. These processes are then adapted dependent on project scope and needs. A large project may require the full suite of PMO services, while a small task can cut the process steps to the bare minimum required to get the job done.

Level 4 organizations gather quantifiable data about their project processes that enables them to "Manage" the processes based on hard data. Quantitative goals are set for processes. This works well for a repeatable process where attributes such as the number of defects or how long a process takes can be measured.

Adopting a proactive analysis and improvement culture is a requirement for an organization to attain Level 5 "Optimization." The effects of process improvement are measured and compared to objectives. Process variation is eliminated. Even if this level is not formally attained via progression through the previous four levels, much can be gained by doing postmortem analysis on projects, with the goal of improving the efficiency of engineering and implementation processes.

Agile programming

Agile programming (AP) and its offshoot "extreme programming," which we'll address later, are two software development methods that address the issue of changing requirements. The manifesto for agile software development is stated simply in four precepts:

- Individuals and interactions over processes and tools;
- Working software over comprehensive documentation;

- Customer collaboration over contract negotiation;
- Responding to change over following a plan.

This is in contrast to traditional, waterfall project methodology, where each step is clearly defined and completed before the next step begins—essentially a linear process. Any major change in requirements or unanticipated delays can wreck the project schedule.

In an AP environment, software projects are done in small, incremental steps. This enables new requirements and plans for future implementation to be integrated into the development process before a point of no return is reached. When a light-weight development methodology is used, value and innovation can be maximized.

Development is modular. Feature-driven development (FDD) is a practice where a product is delivered feature by feature over the life of the project. A useable release is completed and thoroughly tested before moving on to the next feature module. These small victories bolster team confidence and enthusiasm.

In this practice, communication among all stakeholders is ongoing; little emphasis is put on the strict adherence to a formal, staged process; and documentation is a work in progress, carried on by technical writers in parallel with development.

Team selection is also crucial. Highly talented players are imperative; and short, regular meetings are held daily or weekly, including regular meetings with the customer. Time-outs are taken to review progress and plan for new directions.

The project manager is really a coach and is responsible for facilitating communication among the players. Because all the players are "stars," facilitating team collaboration is at the heart of this practice.

Extreme programming

One agile methodology that has become something of a buzzword is extreme programming (XP). XP is intended to create a workflow that can adapt to rapidly changing user requirements.

XP is characterized by rapid feedback, assuming simplicity (there is always an elegant solution to any design task), applying incremental changes, and embracing design changes. Development is broken into iterations based on a small number of features. Iteration cycle time is measured in weeks.

A customer (or customer proxy) is on-site and always available. This is a person who will use the system. This enables constant communication of changing user requirements as well as feedback on feature functionality.

Activities include coding, testing, listening (programmers know more about the desired features than the business aspects), and designing (creating a design structure that logically organizes the system).

Extreme practices

When implementing extreme development methods, it is important to keep the rules and practices of XP in mind.

Fine-scale feedback

- Pair programming—two programmers at one machine; code is reviewed as it is written;

- Planning:
  - Business provides list of features and required functionality,
  - User stories are written for each feature,
  - Estimate of development team's resources is made,
  - Business decides what stories to implement and the priority (order) of development;

- Test-driven development:
  - Unit tests are written by developers to test functionality while the code is being written and automated,
  - Acceptance tests (functional tests) are customer-specified, full-system scripts of user-interface actions.

Continuous process

- Continuous integration—changes are integrated daily;
- Design improvement;
- Small releases start with minimal core functionality, add feature(s) each iteration.

Shared understanding

- Coding standards—all code is written to the same standard, e.g., format;
- Collective code ownership—any developer can work on any part of the code at any time;
- Simple design—find the simplest (most elegant) solution and eliminate all extraneous code that doesn't implement the current feature set.

Programmer welfare

- Sustainable pace—40-hour work week that can be extended to one week of overtime, after which the project must be reevaluated.

The full list is available at http://www.extremeprogramming.org/rules.html.

How Extreme is too Extreme?

Allowing any person to work on any part of any design drawing, database, or other project element in a completely free environment can wreak havoc with the engineering process. This is one attribute of XP that doesn't translate well to any task with more than two people; communication and repetition of effort become issues.

Moving targets

AP and XP attempt to resolve the challenges of changing user requirements for an in-progress project. They employ a development process that can respond rapidly to project modifications without disruption, or extensive rework or redesign.

Documentation is dynamic and tracks code development. It is not unusual for a technical writer to be part of an AP or XP project team. This leaves the application programmers to spend their time on code development.

Both processes are dependent on highly talented designers and communication. A siloed culture will fail at attempting AP or XP. Similarly, the customer must be actively involved and not wait to introduce new or modified features.

With its constantly changing delivery opportunities, technology, and business models, broadcast engineering would seem to be a prime candidate for application of at least some aspects of AP and XP. But many broadcasters believe traditional CMMI and waterfall techniques are better suited to the mission critical 100% reliability necessary for broadcasting, where a small failure can have a large business impact.

Applying CMMI and Six Sigma to broadcast engineering

CMMI techniques can be applied to the overall design and implementation cycle. All engineering departments can benefit from design reviews, peer input, and customer review. Candidates for process formalization include the following:

- Problem resolution: establishing a defined procedure of notification, escalation, resolution, and documentation;
- Interdepartmental projects: establishing how to gather user requirements, develop system documentation, peer design review, and customer approval.

The widely used Six Sigma approach can also be applied to production processes. It's a challenging task, because of Six Sigma's reliance upon quantifiable results, but there are a few potential areas where efficiency can be improved:

- Graphics: Can the design approval process be shortened? Can GFX be repurposed easily for different delivery channels?
- Editing: How long does it take for content to get from the field onto an editing system? Can search time for a clip be reduced?

Each broadcast organization is unique and does things in its own way, but all share a common challenge: Workflows and engineering processes must transition to digital alongside infrastructure technology. Following CMMI and Six Sigma is a way to manage this transition within an accepted framework.

Don't lose sight of the fact that this is about the use and application of technology, not simply process management. An experienced engineer will understand the reasons behind CMMI and Six Sigma, while an "outsider" will see only the process.

There are always opportunities to improve operations in a broadcast center. The engineering of integrated technology media systems provides the opportunity to correct the mistakes and bad habits of the past. Regardless of whether it is achieved via CMMI or Six Sigma, reliable 24/7 on-air integrity is one of the most important aspects of broadcasting.

Time is money: a numerical analysis

Maximizing the ROI on creative tools and minimizing the time spent in each portion of the creative process will allow production teams to meet the increasing demands of multiplatform distribution and consumption. Yet many a broadcaster has upgraded to digital resources while maintaining its analog technology-based workflows.

In the content-creation process, it is important to integrate technology with the workflow. For example, an investment in a graphics media asset management (MAM) system may require a $50,000 purchase. Management may not see the benefit of installing a GFX MAM. They feel that graphics designers can continue as they currently do, and navigate a loosely structured directory tree to find a needed graphics element.

However, a workflow analysis may reveal that graphics personnel spend between 10 minutes and four hours searching for archived content. This may be on a PC, over network resources, or, for physical media, on someone's desk. Let's say the average is two hours. Not only does this impact production time, but two hours of a $52,000 per year designer's time is a $50 expense. If each designer makes 100 searches in a year and you have 10 designers, that's $50,000 a year of time spent searching for graphics content. The MAM pays for itself in one year.

In addition, the two-hour average search time is reduced to five minutes. This results in reducing the production process by nearly two hours. This time can be used to create more graphics (and more complicated graphics). In a multiplatform environment, this time can be used to optimally format GFX content to its delivery methodology and consumption device.

ADAPTIVE MEDIA SYSTEMS ENGINEERING

By appropriately applying both waterfall model and AP project management paradigms, an adaptive project management process can be developed and applied when engineering integrated technology broadcast media systems.

Using a capability matrix model to assess process efficiency implies a traditional waterfall project methodology. Each step is clearly defined and completed before the next step begins; this is an essentially linear process. Any major change in requirements or unanticipated delays can wreck the project schedule.

At the other end of the spectrum, AP methods address the reality of rapidly changing user requirements inherent in software development. Yet in this apparent chaos there is a clearly defined set of principles and practices that are comprehensively adhered to. The key to the agile methodology is short, iterative design cycles, "star" team members, and continuous communication with the client.

A combined methodology

Can two disparate project management methods be reconciled in modern broadcast engineering, selecting the best of each? Just as IT and traditional video engineering are producing a new technical discipline of media system engineering, so too can traditional waterfall project management and AP techniques forge a new project management methodology.

In fact, this is exactly what media system engineering requires for integrated technology systems to be successfully and robustly designed, deployed, commissioned, and maintained. By appropriately applying the processes of both paradigms, an adaptive model can be applied to engineering broadcast media systems.

Determining what parts of a project should be managed by waterfall techniques and which lend themselves to agile methods will not always be easy. Risk management is always a part of good project management, and can help with the decision of which methodology to use where. Applying risk evaluation can instill organizational confidence in an adaptive approach to broadcast media systems engineering.

A scenario

Consider this scenario: A broadcaster has decided to consolidate all its graphics resources into one centralized location. This includes a SAN, GFX render farm, SDI routing to multiple displays, and new workstations and applications—a multimillion-dollar project that will take nine months to design, construct, and commission.

When the design begins, the project scope requires that the GFX produced will only be used in broadcast programming. The budget and schedule are approved, equipment is specified, and construction is scheduled. Three months elapse and everything is on target and on budget.

The organization also maintains a Web site and has moved into repurposed content delivery to mobile/handheld devices. Executive marketing management has decided that consistency of GFX images across all delivery channels will ensure brand awareness among consumers. They are faced with two options: develop an independent GFX production process for each channel, or transcode and reformat GFX for each channel on the fly.

Management communicates to engineering that one GFX process should support all channels, and because a new GFX workspace and infrastructure project is underway, it should be adapted to meet the new multiplatform delivery requirements. There is also one more constraint—the project schedule and budget cannot be altered.

This is exactly the scenario that calls for both waterfall and agile project management. If a systems engineering department follows a formalized design and implementation process, they are now faced with abandoning this ordered methodology and will use ad hoc procedures to get the job done. If agile broadcast system engineering is practiced, however, the engineers could successfully respond to the change in project requirements while designing a solid foundation infrastructure.

Agile broadcast engineering

Because of the rapidly changing production and distribution requirements in today's evolving broadcast infrastructure, some will assert that the traditional engineering process is too slow. With the emergence of new delivery channels, consumption devices and programming formats, projects that take more than a few months to deploy will miss their windows of opportunity—first to market has value.

Software application designers have faced this challenge and developed methods to address dynamically changing requirements, presented during application development. To the uninitiated observer, these methods may seem haphazard and ad hoc, but there is an underlying design process philosophy to both AP and its more radical offshoot, XP.

A natural selection

Upon analysis, the design of a digital broadcast infrastructure can be logically divided into areas where either the traditional waterfall or AP method is the best fit. For example, the SDI distribution network is a stable technology and a long-term investment. Any error in design or implementation will be difficult to correct later. Once installed, the system will only require expansion, not a complete redesign. This is an area where the waterfall design process is a natural fit.

The evolution to 3 Gbps distribution can be accomplished as an upgrade, and although it may not be as short as weeks or months, it can still be thought of as an AP design iteration. How will going to 3 Gbps impact the entire production and broadcast operation; how can this be leveraged to improve efficiency? The Six Sigma analysis process can identify workflow refinements that will best leverage the 3 Gbps infrastructure to improve operational efficiency.

Or, if a team is designing a media asset management application, every new audio or video codec requires a new software module. As users get acquainted with the GUI, in all likelihood they will suggest new features that need to be incorporated. An AP project management process fits best here. As the project management team traverses the "rapids," contingency plans can be put in place that take an agile approach in response to changing requirements. In other words, plan ahead for change.

In general, waterfall techniques fit best with platform and infrastructure projects, while agile methods work well with features. This roughly follows a hardware/software system division, though AP can be applied to a hardware-based project as well. For instance, monitors may be needed for a control room, and the rapid improvements in monitor technology combined with dropping prices will likely result in a selected model soon being eclipsed by a later design. So, it may be best to practice just-in-time procurement, taking into account order delivery lead time. But there is a risk in that, if delivery is late, your control room may be completed and waiting for the monitors.

To make the adaptive approach work, broadcasters and engineers must stay ahead of the technology curve. Participation in technical standards activity, attendance at conferences where emerging technology is discussed, and an ongoing dialog with vendors will enable an organization to better judge what will be available in the future, and how to stay adaptable enough to benefit from it.

ORGANIZATIONAL EVOLUTION

Just as analog workflows don't cut it in a digital world, the organizational structures and methodologies that were used in the broadcast technology domain come up short when the systems include new IT.

Project management and teamwork

Project planning and management are now of primary importance. A timeline and budget is not sufficient. How does this project fit in the overall infrastructure?

The challenge becomes identifying qualified personnel as potential technical project managers. Being technology-dependent projects, it is necessary to have an engineering person take the lead. This individual must have engineering experience and problem-solving skills, be able to bring the proper technologists together to build an effective team, gather and define user requirements, and then guide the team effectively so as to arrive at the best technical solution.

The project team should include representatives from all impacted stakeholders (see Figure 6.4). This will include engineering, IT, production, operations, and

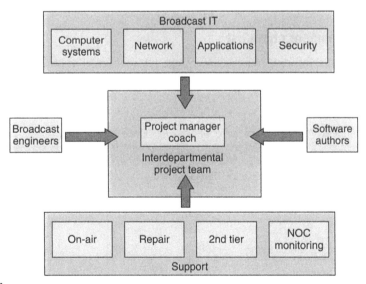

FIGURE 6.4

An interdepartmental project team

creative personnel. All should aid in the design, adding expertise and information pertinent to make the infrastructure monitorable and facilitate troubleshooting.

It is important to include support personnel in the earliest design phases. Front-line, first responder experience, incorporated into the first stages of system design, will save countless hours of problem resolution in the future. The BOC will be "designed for support."

TECHNOLOGY—LIFE CYCLE MATRIX MODEL

In order to understand the interdependence of all components affected by the transition to digital, let us consider the premise that the media business, in general, can be modeled as a two-dimensional matrix. In the vertical direction, the technology axis has as its foundation in the physical, and then ascends to the media network, applications, and security layers. Each of these technologies has an influence on the four phases of the horizontal axis, progressing over time from creation through assembly, distribution, and then consumption. Tables 6.1 and 6.2 illustrate the media technology–life cycle matrix (MTLM) concept, comparing analog NTSC and digital ATSC.

Understand that the matrix is merely a type of technology by function checklist. The idea is that as a system is underdevelopment, rather than lump the PC, storage, network and applications all under one task, by breaking them out into individual areas of expertise and approaching each as a system subtask, a more efficient and reliable implementation will be attained. Reliability is accomplished by paying attention to the details; after all, that's where the design devils live!

Design development checklist

A new engineering discipline is emerging. Technology is in its infancy. Media, in particular HD video, is extremely demanding, and pushes technology to its limit and

Table 6.1 Simplified Analog TV Media Technology/Life Cycle Matrix

	Creation	Assembly	Distribution	Consumption
Security	Badge access	Badge access	FCC license	On/off switch
Application	Manual	Manual switching	VSB	Entertainment
Media network	Distribution amplifiers Signature routing	PCR MCR	Over the air	Antenna
Physical	Audio Video Graphics	Commercial insertion Bugs Logo	CVBS	TV receiver

beyond. Each of the four layers requires expertise gained after years of school, training, and real world experience.

TRANSITION TO MEDIA SYSTEMS ENGINEERING

By now, most broadcast engineering departments have accepted the influx of IT and the dependence on computers as a fact of life in the digital age. With this increase in system complexity, the cross-functional, interdepartmental teams are required to design, install, commission, and maintain the digital infrastructure. Given the fact that system complexity and technical expertise is beyond the ability for a single person to master, a new emphasis on teamwork and process efficiency has emerged.

Many broadcast engineers have made valiant efforts to gain expertise in IT and computer science. Conversely, those in the IT domain, called to work with broadcast systems, have had to learn the ways of broadcast engineering. But as Gavin Schultz said at the 2004 Pasadena Engineering Conference during his tenure as president of

Table 6.2 Comprehensive Digital ATSC Media Technology/Life Cycle Model				
	Creation	**Assembly**	**Distribution**	**Consumption**
Security	User profiles AAF DRM EAS	IDS/IPS MXF DRM EAS	Encryption Broadcast flag V-ISAN DRM EAS	Copy protection NDIIPP DRM EAS CAS
Application	Graphics Animation Audio	Automation Asset management	DASE OCAP/ACAP VOD PPV	Entertainment ITV Games MP3 Program guide MAM/search
Media network	Network Storage Backhaul Contribution Ingest	Network Storage PCR MCR Playout	Terrestrial Cable Satellite Internet Cellular Radio	Personal home digital network
Physical	Audio Video Graphics	MPEG transport stream Commercial insertion Bugs Logo	8-VSB QAM CODFM IP	TV Internet DVD Audio Digital cinema

the SMPTE, neither broadcast engineering nor IT is absorbing the other; in reality, a new engineering discipline is being born.

The rate of acceleration is accelerating!

The rate of change of all aspects of the broadcast industry—business, content and technology—is changing at an ever-increasing rate. In fact, the rate of acceleration of the transition is accelerating—a positive third derivative! The scope has expanded beyond solely television, and it may be a decade before disruption stabilizes. How do broadcasters adapt?

Disruptive transition of this magnitude challenges broadcasters to ensure that they are not backed into a corner. They must anticipate all potential revenue models. It is not a question of implementing IPTV, content download, or targeted advertising, but of how to engineer the infrastructure and production processes to optimize cost and time efficiently, and thereby derive the maximum profit from all delivery channels.

The rest of this book will present various aspects of the four phases of the multiplatform media life cycle with respect to the four infrastructure layers. As technologies merge, platforms multiply and the audience takes control of a personalized media experience; the chapters that follow will strive to bring the big picture into focus by shedding light on how this affects the broadcast industry.

Content distribution networks

From the invention of telegraph and the start of electrical communications, distribution channel technology has governed the amount of information that could be transmitted in a given time interval. No matter how one tries, the physics of the distribution channel determine how much information can fit in the pipe.

Bandwidth is a fundamental characteristic of a communications channel. It is a measure of data transfer capacity. Bits per second metrics are used for serial, single-wire paths; bytes per second metrics are used for parallel data transfers; and symbols per second metrics are used for more sophisticated methods of digital transmission.

Content distribution networks can be broken down into two primary categories: consumer and commercial. Consumers are end users receiving content wirelessly over the air and over a wired physical distribution technology. Each method has its pros and cons as well as implementation suitability scenarios. Content distribution to consumers will be covered in Book 2.

Content producers and broadcasters use sophisticated distribution networks that ensure maintenance of the highest possible audio and video quality. Professional content delivery networks are used in three phases of the content life cycle.

During content creation, electronic newsgathering (ENG) trucks or remote venues feed content back to an operations center. This backhaul acquisition of content is assembled into a format for distribution. However, disseminating the content can be done in a number of ways.

If the production and transmission site are colocated, signals can be released over the distribution channel. This is rarely the case. Consider a radio broadcaster located in a city. The program originates from a studio location, but the transmitter is in the suburbs. The program must now be sent to the transmitter over a studio transmitter link (STL) that uses technology that is incompatible with consumer reception.

Similarly, a TV network releases programming to affiliates over dedicated commercial transmission paths. Local commercials are inserted at each of the affiliate stations. When sending an e-mail, the signal on the cable to your computer hits the ISP gateway and traverses over the Internet's commercial backbone until it reaches a mail server on a network that can be accessed by the recipient. There the message is downloaded over a dedicated connection to the recipient's computer.

Before delving into content production, an understanding of the technology and limitations of each channel will explain constraints on creating and assembling content for each delivery method and consumer platform.

DISTRIBUTION SCENARIOS

Transporting broadcast quality digital content presents unprecedented challenges. With analog signals noise, signal strength and reflections were of paramount concern. Channel bandwidth only needed to support baseband signals: 6 MHz for NTSC system and 8 MHz for PAL and SECAM.

Compressed video and audio requires attention to a new set of technical issues. Since compression is usually lossy and perceptual based, as the bit rate is reduced, increasingly more high-frequency information is lost. As low bit-rate content travels through the production workflow, additional decompression/compression cycles will discard more information. Eventually, audible and visible artifacts will be perceived.

To mitigate the loss of information, higher bit-rate audio and video formats are used in transferring content to, from, and between broadcast sites and operations centers. Low bit rates, 19.39 Mbps for ATSC DTV, are used only for finished content meant for consumer distribution.

Backhaul and contribution feeds from remote venues use bit rates from 38 Mbps up to 220 Mbps for HD and 40 Mbps for SD video content. Distribution for network release may use a mezzanine level of compression, typically 45 Mbps. The data rate is dependent on the delivery channel, and cannot exceed the lowest bandwidth anywhere on the transmission path.

An STL is used to transfer finished programs to the transmission site; a 19.39 Mbps MPEG-2 transport stream over an ASI or SMPTE 310 interface is sufficient.

Figure 7.1 illustrates how content is distributed to and from segments of the broadcast chain.

Terms describing different areas of content distribution tend to be used loosely. Contribution and backhaul are sometimes used interchangeably; so are distribution, release, and mezzanine.

This chapter will discuss the technology used to backhaul, contribute, and distribute high bit-rate content between operation centers, transmission site, and consumer distribution nodes.

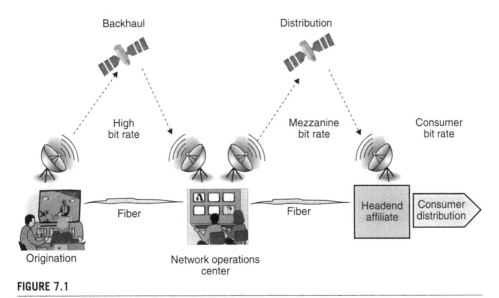

FIGURE 7.1

Distribution technology interdependence during the content life cycle

THE ORIGINS OF TELEVISION LONG-HAUL DISTRIBUTION

AT&T demonstrated long-distance transmission of television in 1927. An experimental electromechanical system scanned 50 lines of resolution at 16 times per second and transmitted the signal over telephone lines from Washington, DC to Whippany, NJ, and then over a radio link to New York. This established the long-haul paradigm that was to remain in place for TV signal long haul until the 1970s: a combination of RF and cable technologies.

Researchers at Bell Labs patented coaxial cable in 1929. Although developed to increase telephone transmission capacity, it turned out to be well suited for long-distance TV transmission. It had an astounding bandwidth for its time—1 MHz, which was a great improvement over unshielded cabling bandwidth. By 1936, experimental coaxial cable was run between New York and Philadelphia, with signal booster stations every 10 miles (16 km).

After this experiment, AT&T discontinued work on electromechanical television, and as conventional industry intelligence realized, the future of TV would be an all-electronic system. Prototype systems were invented and constructed independently by Vladimir Zworykin and Philo T. Farnsworth. The first electronic television system to go on the air was the 405-line UK system in 1936 developed by John Logie Baird.

Bell Labs gave demonstrations of the New York–Philadelphia television link in 1940-1941. AT&T used the coaxial link to transmit the Republican National Convention in June 1940 from Philadelphia to New York City, where it was televised to a few hundred receivers over the NBC station.

NBC had earlier demonstrated an intercity television broadcast on February 1, 1940, from its station in New York City to WRGB in Schenectady, NY using General Electric relay antennas, and began transmitting some programs on an irregular basis to Philadelphia and Schenectady in 1941. Wartime priorities suspended the manufacture of television and radio equipment for civilian use from April 1, 1942 to October 1, 1945.

In 1946, AT&T completed a 225-mile (362 km) cable between New York City and Washington, DC. The DuMont Television Network, which had begun experimental broadcasts before the war, launched what *Newsweek* called "the country's first permanent commercial television network" that same year, connecting New York with Washington, DC. Not to be outdone, NBC launched what it called "the world's first regularly operating television network" on June 27, 1947, serving New York, Philadelphia, Schenectady, and Washington, DC.

In the 1940s, the term "chain broadcasting" was used, as the stations were linked together in long chains along the east coast. But as the networks expanded westward, the interconnected stations formed networks of connected affiliate stations.

When television broadcasting began in earnest after the conclusion of WWII, AT&T began providing transmission for broadcasters over cable between Washington, DC., and New York. The initial distribution network was expanded by the addition of a microwave-relay transmission system between New York and Boston in 1947. When it was connected to the New York–Washington cable, television networking from Boston to Washington, DC was possible.

In 1951, AT&T completed the construction of their microwave radio relay network of AT&T Long Lines, the first transcontinental broadband-communications network, and carried all four existing TV networks coast to coast. President Harry Truman's September 4 address to the United Nations/Japan Peace Treaty Conference is the first live transcontinental television broadcast.

These networks were the only way to distribute TV programs nationally until two technologies developed in the 1960s and commercialized in the 1970s revolutionized long-haul television distribution: satellites and fiber optics.

DISTRIBUTION TECHNOLOGIES

Distribution technology can be divided into two categories: those that are freely propagated over a medium and those that are carried over some kind of physical conveyance. Radio frequency transmission is an example of the former, while cable-based transmission is an example of the latter.

RF transmission is over the air, or via satellite. Cable infrastructures are either wire-based or optical. There is one exception: for high power, RF wave guides are used, but these are beyond the scope of this discussion.

Radio wave propagation in space

Radio waves are electromagnetic waves that propagate through space. There are three types of radio waves: ground, space, and sky. Figure 7.2 illustrates the different transmission characteristics of each type.

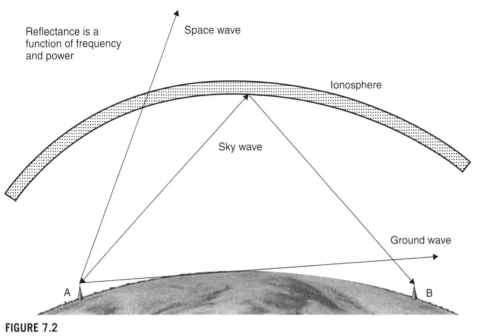

FIGURE 7.2

Radio wave propagation through free space

Consider what happens for each type of wave when transmitted from point A. A ground wave is eradiated in all directions from the transmitter. A ground wave will not be received at station B because of the curvature of the earth. A space wave travels through the ionosphere and is usually a concentrated, directional beam. Sky waves bounce off the ionosphere and return to earth over a greater distance than ground waves. In this example, they are the only type of the three that can facilitate communication between point A and point B.

Microwave

Microwave transmission refers to the technique of transmitting information over a microwave link. Microwaves are electromagnetic waves with wavelengths shorter than one meter and longer than one millimeter, or frequencies between 300 MHz and 300 GHz. Due to the fact that signals in this frequency range are affected by rain, vapor, dust, snow, cloud, mist, and fog, the successful implementation of microwaves is generally limited to line-of-sight transmission links without obstacles.

Because a line-of-sight radio link is made, the radio frequencies used occupy only a narrow path between stations. Antennas used must be highly directional; these antennas are installed in elevated locations such as large radio towers in order to be able to transmit across long distances.

Typical types of antenna used in radio relay link installations are parabolic reflectors, shell antennas, and horn radiators, which have a diameter of up to four meters.

Highly directive antennas permit an economical use of the available frequency spectrum, despite long transmission distances.

Satellite

Placing an artificial moon in orbit around the Earth had long been a dream of many a scientist. Isaac Newton first revealed the physics and mathematical principles required to do it, although it could not be accomplished until more than two centuries after his death.

Arthur C. Clarke described the concept of a stationary orbiting object in October 1945 in a paper titled *Extra-Terrestrial Relays—Can Rocket Stations Give Worldwide Radio Coverage?*, published in *Wireless World*. The geostationary orbit is a special case that occurs when a satellite's angular velocity matches the Earth's. The object in orbit will remain in a fixed position relative to the surface of the Earth.

In 1957, the Soviet Union launched Sputnik. While the world was in awe, the U.S. industrial–military complex was in shock. Military leaders understood the strategic importance surveillance and communication orbiting satellites would have, especially with further development and refinement. In addition, the ability to deliver a nuclear weapon via intercontinental ballistic missile (ICBM) was closer to fact than to fantasy during the Cold War. The United States had to be in space.

Launch failures by the United States were forgotten when Explorer I made it to orbit in January 1958. While Sputnik simply transmitted a periodic radio beep (at a frequency that could be picked up by ham radio operators) back to the Earth, Explorer was packed with scientific instruments. When the telemetry transmitted back to the Earth was deciphered, radiation belts around the Earth were discovered and named after James Van Allen, a professor at Ohio State who designed the on-board instruments for the Jet Propulsion Laboratory.

Communication via space

In the 1950s, J. R. Pierce and his associates at Bell Labs began work on satellite communications concepts. At that time the responsibility for all communications satellites rested with the military.

The National Aeronautics and Space Administration (NASA) was established in July 1958, by the National Aeronautics and Space Act. Two years later, Congress defined the scope of civilian and military space efforts: NASA would conduct research with "mirrors" or "passive" communications satellites, while synchronous and "repeater" or "active" (satellites that amplify the received signal at the satellite) satellites would be a Department of Defense effort.

Echo 1A, a metallic balloon, was placed in low Earth orbit (LEO) by a Delta launch vehicle in August 1960. The first communications satellite "Echo" could be seen with the naked eye.

In 1960, AT&T filed with the Federal Communications Commission (FCC) for permission to launch an experimental communications satellite. By the fall of 1960, AT&T began development of a satellite communications system called Telstar.

The operational system would consist of "between 50 and 120 simple active satellites in orbits about 7,000 miles high."

The Influence of the Space Program on Digital Technology

Soviet rockets could carry a considerably heavier payload than U.S. rockets. Due to this "missile gap," great importance was placed on the need to miniaturize electronic systems. As a national defense priority, refinement of integrated circuit (IC) fabrication progressed rapidly. At one point in the 1960s, it is said that 60% of all the ICs produced were for the space program. The missile gap disappeared with the development of the F1 engine and the Saturn moon rockets.

It's difficult to hit a moving target

A satellite in orbit travels at over 17,800 mph; therefore, the amount of time in which a signal can be sent from point A to point B over a satellite link is limited to minutes. This is why Bell Labs envisions a network of orbiting communications satellites. Using cellular concepts, signals would be handed off to a trailing satellite when the satellite currently broadcasting the signal moved out of range of the ground stations.

A better solution was needed if the use of space for dependable communication had to be viable and commercially profitable.

Deployment accelerates

Bell Telephone Laboratories continued its pioneering work and designed and built the Telstar spacecraft. The first Telstars were prototypes that would prove the concepts behind the constellation system that was being planned. NASA's contribution to the project was to launch the satellites and provide some tracking and telemetry functions, but AT&T bore all the costs of the project, reimbursing NASA $6 million.

Telstar I was launched on July 10, 1962, and once in orbit, it immediately broadcast live television pictures of a flag outside its ground station in Andover to France. The first "official" transmissions occurred on July 23, when a short segment of a televised major league baseball game between the Philadelphia Phillies and the Chicago Cubs at Wrigley Field was transmitted to Pleumeur-Bodou, France. On that day the first satellite telephone call, faxes, and data transmissions were also performed.

The first geosynchronous communication satellite, Syncom 2, was launched in 1963, and by 1964, two Telstars, two Relays, and two Syncoms were operating successfully in space. During this time, the Communications Satellite Corporation (COMSAT), formed as a result of the Communications Satellite Act of 1962, was contracting for their first satellite. COMSAT ultimately chose geosynchronous satellites proposed by Hughes Aircraft Company for their first two systems and TRW for the third. On April 6, 1965, COMSAT's first satellite, also the world's first commercial communication satellite, Intelsat I (also called Early Bird), was launched from Cape Canaveral.

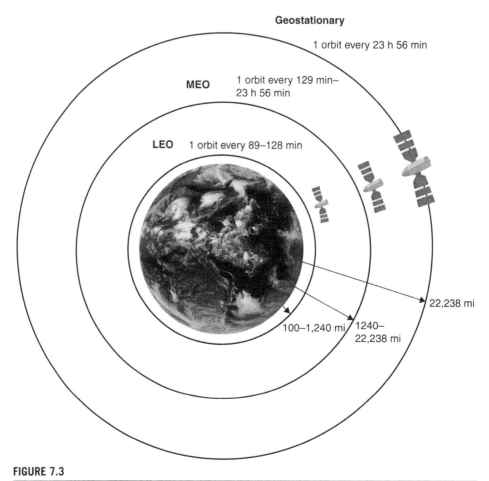

FIGURE 7.3

Low, medium, and geostationary orbits

Satellites can be deployed at various altitudes above the Earth: LEO, medium Earth orbits (MEO), and geosynchronous orbits, which have become the de facto orbit for communication satellites. Figure 7.3 compares orbital types.

Today satellites are an integral component of content and data distribution. However, rather than remaining strictly divided, satellite, cable, and terrestrial transmission together form content distribution networks.

Superstations pave the way

Cable TV systems were little more than local aberrations in the broadcast universe until satellites were brought into the picture. In the late 1970s, Ted Turner had the idea of distributing WTCG in Atlanta via C-band to cable systems across the United States. Now available across the country, WTCG (later changed to WTBS and now WPCH) became the first "superstation."

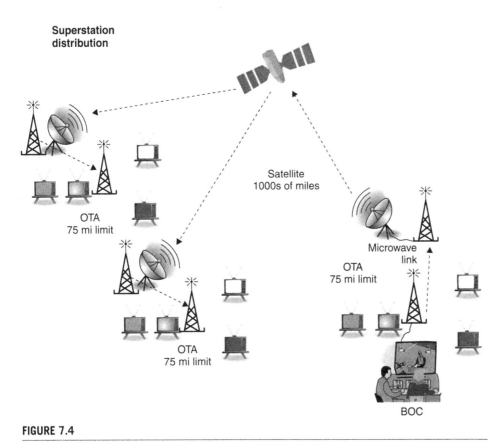

FIGURE 7.4

Superstation national television distribution

Turner's distribution method changed the rules of the game. FCC Rules and Regulations in 47 CFR 77.120 define a "nationally distributed superstation." The general requirement is that the station is "not owned or operated by or affiliated with a television network," offers "interconnected program service on a regular basis for 15 or more hours per week to at least 25 affiliated television licensees in 10 or more states," and is retransmitted by a satellite carrier. Figure 7.4 illustrates the long-distance distribution technique.

The significance of this conceptual innovation was that when Cable News Network, CNN, was launched in 1981, Turner used the same national content distribution paradigm: satellite uplink and national distribution via downlink to cable headends. So, in a very real sense, superstations morphed into national cable TV broadcasting systems.

AT&T innovates

AT&T launched its first Telstar III domestic communications satellite in 1983. This resulted in a short-lived attempt to introduce direct-to-home satellite broadcasting.

Japanese television had already implemented such a system and planned to use it to deliver its MUSE HDTV system.

AT&T continued to expand its commercial television contribution and distribution network. Optical networks and satellite systems were integrated. Its Littleton, CO-based operations center, Headend in the Sky (HITS), became a major distributor of content to MSOs, delivering more than 170 digitally compressed video and audio TV programming signals to 3,000-plus cable operation sites across the United States. AT&T got out of the satellite business in 1994 when it sold the fleet to Loral. In 2001, Comcast acquired AT&T Broadband, HITS, and AT&T's cable TV business.

Satellite transmission technology

Satellite television systems consist of Earth and space segments. The transmitting antenna is located at an uplink facility. Uplink satellite dishes are very large, as much as 9–12 meters (30–40 feet) in diameter. Transmission is in the C-band or K_u-band or both. A high band, the K_a-band, is being increasingly utilized. Table 7.1 lists the uplink and downlink frequencies for each band.

An uplink dish is pointed toward a specific satellite, and the uplinked signals are transmitted within a particular frequency in a band and are received by one of the transponders tuned to that frequency aboard that satellite.

The transponder "retransmits" the signals back to the Earth at a different frequency band (a process known as translation that is used to avoid interference with the uplink signal). The leg of the signal path from the satellite to the receiving Earth station is called the downlink.

Satellites have up to 32 transponders for K_u-band or 24 for a C-band. Typical transponders each have a bandwidth between 27 MHz and 50 MHz. C-band transmission is susceptible to terrestrial interference, while K_u-band transmission is affected by atmospheric moisture.

Each geostationary C-band satellite needs to be spaced 2° from the next satellite to avoid interference. For K_u the spacing can be 1°. This places a limit on the total number of geostationary C-band and K_u-band satellites at 180 and 360, respectively.

The power of a radio transmission decreases in proportion to the square of the distance it travels. A weak downlinked signal is reflected by a parabolic receiving

Table 7.1 Satellite Frequency Bands

Frequency band	Downlink	Uplink
C	3,700–4,200 MHz	5,295–6,425 MHz
K_u	11.7–12.2 GHz	14.0–14.5 GHz
K_a	17.1–21.2 GHz	27.5–31.0 GHz

dish to a feedhorn at the dish's focal point. This feedhorn is the flared front end of a section of waveguide that "conducts" signals to a probe or pickup connected to a low-noise block (LNB) downconverter. The LNB amplifies the weak signals, filters the block of frequencies in which the satellite TV signals are transmitted, and converts the block of frequencies to a lower frequency range in the L-band range, 950–1450 MHz.

The satellite receiver demodulates and converts the signals to television, audio, or data. If the receiver includes the capability to unscramble or decrypt, it is called an integrated receiver/decoder (IRD). The cable connecting the receiver to the LNBF or LNB should be of low loss type, such as RG-6 or RG-11.

Fiber-optic communication

Fiber-optic systems consist of a light transmitter (laser or LED); an optical medium (glass or composite); and a photodetector receiver.

The first prototype laser (an acronym for Light Amplification by Stimulated Emission of Radiation), was a synthetic ruby crystal device demonstrated in 1960 by Hughes Research Laboratories. Although commercial lasers became available in 1968, it took refinements to glass purity by Corning and the development of room-temperature operational semiconductor diode lasers at Bell Labs in the early 1970s to make fiber optics a viable commercial technology.

A variety of light-emitting technologies are used in transmitters. Fabry-Perot lasers or distributed feedback (DFB) lasers are used in long haul and high data rate applications. Vertical-cavity surface-emitting lasers (VCSELs) are suitable for shorter-range applications such as Gigabit Ethernet (GigE) and Fibre Channel. Light-emitting diodes (LEDs) are used for short-to-moderate transmission distances. LEDs are the least-expensive transmitters but have limited data capacity.

Figure 7.5 shows the development of, and subsequent improvement in, fiber-optic wavelength windows over the last few decades. Appropriate transmitter technology is also indicated for each window.

Two types of photodetectors, avalanche photodiode (APD) and positive-intrinsic-negative (PIN), convert photons of light to electrons. Because of the small number of photons received, amplification is necessary to recover data and produce a usable signal. APD amplification is internal, while the amplification is external for PIN detectors.

Fiber-optic systems use a variety of signal multiplexing techniques. Time-division multiplex (TDM) assigns data packets to time slots and is used in long-haul infrastructures. Wave-division multiplex (WDM) enables multiple wavelengths of light to share a single fiber. In first-generation deployments, WDM technology supported just two wavelengths, also referred to as "lambdas," usually 1,310 and 1,550 nm.

As fiber-optic technologies improved, it became possible to transmit more than two lambdas simultaneously over a single fiber strand. This resulted in the development of coarse wave-division multiplex (CWDM) and dense wave-division multiplex (DWDM). DWDM uses narrow channel spacing, frequently 0.8 or 1.6 nm, while

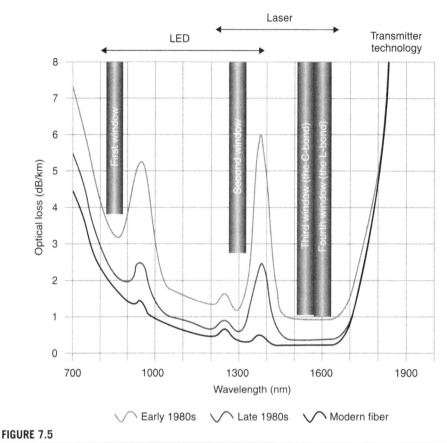

FIGURE 7.5

Improvements in fiber optics

CWDM spaces channels 20 nm apart. Figure 7.6 illustrates implementation of each technique as specified in ITU-T standards.

Single-mode fiber (SMF) carries a single wavelength of light and is suited for long runs, such as between buildings, venues, and broadcast sites (STL, TSL, intracity, and intercity links), for long camera runs and as risers in facilities. SMF cables are yellow. Fiber cores are 8.5 micrometers in diameter. Something of an oxymoron, SMF is best suited for DWDM implementations. This is because DWDMs pack multiple lambdas so tightly that the bundle can be transmitted as a "virtual single" wavelength.

Multimode fiber (MMF) can carry multiple wavelengths on a single strand. They are used in short runs generally inside a building and are identified by their orange color.

SMF technology is more expensive to implement than multimode. Lasers must be precisely tuned, and cannot use the less expensive LEDs transmitters found in CWDM links.

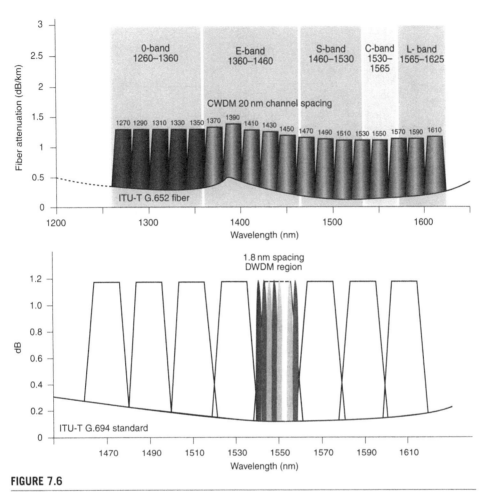

FIGURE 7.6

Comparison of CWDM and DWDM

DIGITAL SIGNALING

Analog television transmission has all but completely disappeared. Digital rules! Digital transmission is enabled by audio and video compression technology. HD data rates of 1.5 Gbps require a 50:1 data reduction to enable them to be carried in a 6 MHz channel.

Digital transmission and reception is an all-or-nothing proposition. Either the signal can be perfectly decoded or virtually not at all. This characteristic is called the cliff effect.

Decoding accuracy rapidly deteriorates in the cliff area: "salt and pepper", macroblocking, and other artifacts become perceptible. Error correction and concealment techniques are implemented to extend the ability of receivers to recover and decode corrupted digital signals.

Synchronous

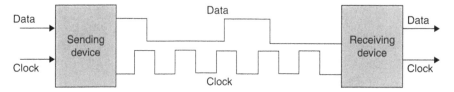

Asynchronous

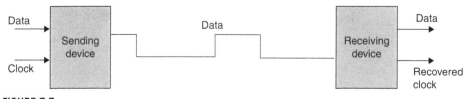

FIGURE 7.7

Synchronous and asynchronous data transmission

Random, constant, and deterministic

Digital data transfers can be broken down into three categories. The first is synchronous, where data is continuous at a given, uninterrupted rate over time. SDI is a synchronous digital signal.

Asynchronous data transfer techniques can be illustrated by the Internet; packet delivery has no fixed relationship to time. A download can be virtually instant, or can crawl at peak usage times. The packet transfers, and subsequently the number of data bits, have no fixed relationship to a constant duration of time. This is bad for both real-time delivery and media presentation.

Synchronous data transfers, as shown in Figure 7.7, are enabled by transmission of the data along with a clock signal. The receiver uses the clock to identify when data is valid. In an asynchronous system, data is encoded in such a way that the receiver can recover the clock signal and then interpret the information present in the signal to recover the data.

Isochronous data transfer is a technique that delivers a defined quantity of bursty data in a given time duration is called (Figure 7.8). In this way, a device can expect a certain amount of data and needs only to be able to buffer within specified maximum and minimum amounts.

Reception of bursty data can be a problem. Processing circuits and algorithms, especially real-time systems, expect data to be available on a continuous basis. Techniques have been developed to supply data when needed.

Buffering: smoothing out data bursts

In order for a digital processing devices to operate in real time, a defined amount of data must be transferred over a given time period. Buffering is a technique that smooths out bursty data received over a transmission channel.

Nondeterministic

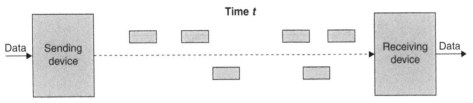

Varying number of packets over a given time interval

Isochronous

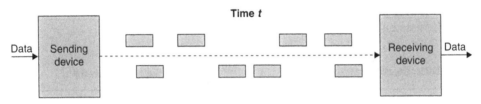

Constant number of packets over a given time interval

FIGURE 7.8

Isochronous communication transmits a constant number of packets over a defined time interval

In order for a buffer to perform its function properly, data must be isochronous. Isochronous data transfers supply a defined amount of data in a given time period, with no specification of maximum or minimum burst size. This specification enables the design of buffers with adequate memory allocation.

A smoothing buffer accepts bursts of data packets and stores them in RAM. This enables access by processing devices at a predictable data rate.

Consider the scenario where a device requires a constant data rate of 100 Mbps and data can burst at 200 and 50 Mbps. As shown in Figure 7.9, data is initially written into a buffer at 100 Mbps for two seconds. Data readout is delayed until after one second when the buffer holds 100 Mb of data. Next, two seconds of 200 Mbps data is fed into the buffer, and at the three-second mark, 300 Mb remain in the buffer (400 Mb have been written into the buffer, while 200 Mb have been read out; add the original 100 Mb and 300 Mb are left in the buffer). The variable data rates are well behaved and can be accommodated as shown.

A buffer smooths variable data rates, enabling a 100 Mbps readout.

If too much data arrives in a given time period, the buffer will overflow and packets will be lost. Figure 7.10a illustrates this situation. The buffer can hold 400 Mb of unread (or unprocessed) data. As in Figure 7.10b, data arrives at 100 Mbps for the first two seconds. After the first second, data readout begins. The data rate now increases to 200 Mbps, and after three more seconds, the buffer is full. During the

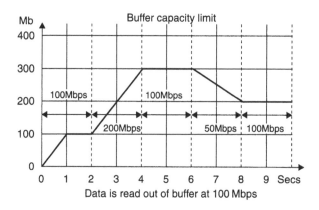

FIGURE 7.9

Buffer fullness as a function of data rate

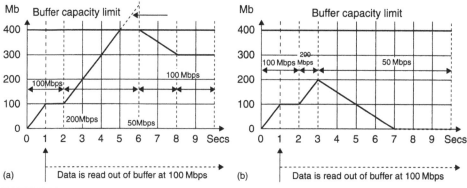

FIGURE 7.10

Buffer overflow and underflow conditions. In (a, left), a buffer overflows and loses data when data arrival exceeds the buffer capacity. In (b, right), a buffer underflows and data is not available for processing when it is read out at 100 Mbps.

next second, half of the data that arrives at 200 Mbps is stored and read out, and the other half is lost. If the data rate now falls to 50 Mbps, data can be read out again at 100 Mbps. After two seconds, in this example, the data rate returns to 100 Mbps and the buffer stabilizes with 300 Mb of its capacity used.

Figure 7.10b follows the same pattern for the initial three seconds as described in Figures 7.10a. After three seconds, the buffer has 200 Mb of data stored. Data now arrives at 50 Mbps, and after four seconds of reading out data at 100 Mbps, the buffer is empty. As data continues to arrive at 50 Mbps, the buffer cannot supply sufficient data to support a 100 Mbps readout.

In the underflow condition, a processing device will be starved for data. This makes real-time device performance (processing) impossible. Conversely, in the overflow condition, the processing device will have too much data to process; hence, it will choke. Real-time processing will continue but with lowered quality, because data will be lost.

BACKHAUL AND CONTRIBUTION

Transporting remotely originating program content back to the studio is referred to as backhaul. Broadcast auxiliary service (BAS) was an early method that got the job done. BAS is still in use today.

Two methods are generally in use for remote broadcasts. One places a production control room, often a truck, at event sites and backhauls produced content back to the operations center for commercial, logo, closed caption, and other content and overlay insertion. The other is an emergent technique that backhauls individual elements to the operations center where audio and video sources, along with graphics, are assembled in a PCR.

Virtually 100% of all TV programming is distributed in a compressed digital form. Each method of transfer has particular technology and content formats associated with it.

Because of the limited distances radio waves can propagate, from the earliest days of radio, programming was distributed using cables over long distances to broadcast operation centers for over-the-air transmission.

Today, because of the data rate demands of contribution-quality HD, content distribution systems require an upgrade to 270 Mbps or greater. Some HD IP video backhaul solutions are based upon JPEG2000 compression, a wavelet technology that enables frame-accurate processing. Using JPEG2000 compression, the bit rate of 1.485 Gbps for HD-SDI can be reduced to between 50 Mbps and 200 Mbps; rates then can be carried over an OC-3 or OC-12 network, respectively.

JPEG2000 is wavelet-based and encodes each frame independently. It provides 10-bit resolution inherent in the HD-SDI format. HD television formats supported are 720p59.94, 1080i59.94, 720p50, and 1080i50. JPEG2000 has been selected as the standard file format for digital cinema. It does not create blocking defects and support resolutions beyond 10-bit HD.

Broadcast auxiliary service

In the early days of broadcasting, production and transmission operations were often in the same location. Content was distributed from the studio to production control rooms over the facility routing infrastructure. This consisted of cables and patchbays, much like existing telephone switchboards. Connecting signals between master control and the transmitter could be accomplished using cabling.

However, as the need to transport signals from locations outside of the operations center increased, BAS was developed. Microwave links were frequently used. Figure 7.11 shows the various BAS scenarios.

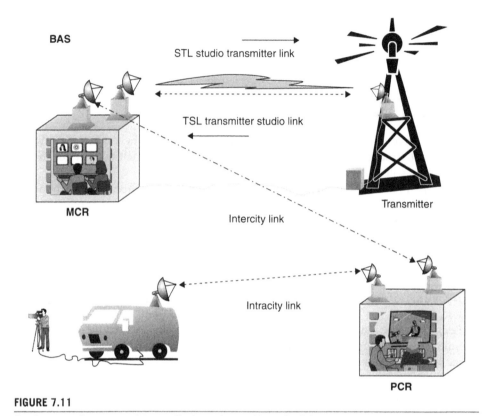

FIGURE 7.11

Studio-transmitter, transmitter-studio, intracity, and intracity links

A microwave BAS link is a simplex or duplex communication circuit between two points utilizing a single frequency/polarization assignment. A duplex communications circuit would require two links, one link in each direction.

BAS frequency bands include 450 MHz, 2 GHz, 7 GHz, 13 GHz, 40 GHz, and the entire broadcast TV spectrum. IFB and two-way microphone communications use the 450 MHz band. The other bands are used for ENG intercity relay (IRC) and in either direction between the studio and the transmitter (STL/TSL).

BAS, is defined in 47 CFR, Parts 101 and 74.

Electronic newsgathering

Coverage of breaking stories is a staple of news broadcasts. Every station in a local market is striving to break a story first or get that one crucial, newsworthy shot on air. BAS transmissions are often used to get the content back to the broadcast center. Usually referred to simply as ENG, electronic newsgathering is a representative example of the transmission chain used by BAS. Figure 7.12 represents the functional blocks of the transmission and receiving equipment.

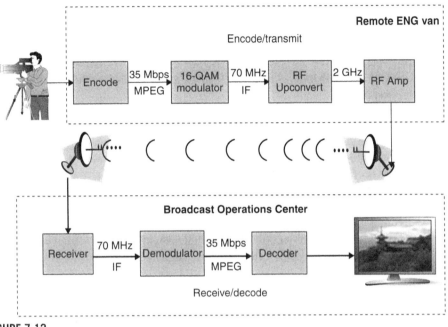

FIGURE 7.12

ENG microwave functional diagram

In this example, the content acquired by the camera operator in the field is compressed using MPEG-2 to a 35 Mbps data rate. This is a high enough rate to allow editing and maintain image quality. 16-QAM modulation produces a 70 MHz intermediate frequency, which is upconverted to a 2 GHz signal for transmission. The dishes are in a line of sight, and at the broadcast operations center, the signal processing is reversed, and the audio and video are ready for editing or live broadcast.

2-GHz BAS relocation

The 2 GHz frequency range is a BAS band primarily used for ENG. Nextel cellular services utilized frequencies in the 800 MHz spectrum range. Unfortunately, this band was also used by public safety licensees. The result was unwanted interference. The FCC solution, reached in 2004, was that Nextel would relinquish its 800 MHz spectrum in return for new allocations in the 2 GHz range. In exchange, Nextel would underwrite the movement of all BAS services. This is known as the 2-GHz BAS relocation plan.

The relocation requires the conversion of each 2 GHz system to a new FCC-mandated frequency allocation and channel plan. This consists of seven digital channels operating at 2,025–2,110 MHz. The new frequency allocation reduced channel widths from 17 MHz to 12 MHz and required digital COFDM transmission. For the first time, the FCC has forbidden analog transmission in a BAS frequency band.

As part of the terms of the relocation agreement, Nextel is required to provide "comparable facilities" to eligible broadcasters for digital operation in the new, reduced bandwidth channel plan. This requires that the existing 2 GHz transmitters, receivers, central receive antennas, and remote controls be upgraded or replaced. Nextel agreed to pay $4 billion to the FCC for the spectrum, minus the cost of the replaced equipment.

Each central receive antenna system requires a wide dynamic range low-noise amplifier (LNA) and a new RF-band pass filter. The dynamic range of the LNA influences the maximum area of coverage for digital operations. A high signal level from an ENG transmitter close to the central receive site can saturate the LNA and/or overload the receiver, which will cause a digital shot to fail. Therefore, the LNA must be able to accept strong signals and it also must have a variable output level. The new RF filter is required to suppress out-of-band interference.

BAS remote control systems need to interface with the new digital receiver-demodulator. Operators will need to monitor RSL as well as other digital signal quality parameters, such as bit error rate (BER), to align the shot and avoid cliff effect link failures.

LONG-HAUL DISTRIBUTION

Content distribution follows a three-layer networking model. Long-haul segments will use either leased commercial network lines or satellite transponders. Dual paths are reserved for backup. Sometimes both satellite and fiber are used—one as primary, the other as backup.

A distinction is made between incoming and outgoing feeds. Incoming is considered backhaul. Outgoing can be divided into contribution and distribution. Contribution is raw material meant for inclusion in a final product. It is usually transmitted using higher quality than consumer transmission, but some scenarios will use consumer-grade or lesser-quality transmission paths, such as satellite phones for ENG.

Distribution is when a final program is transmitted to a consumer-facing system, such as cable systems, broadcast affiliates, and direct broadcast satellite (DBS) systems such as DIRECTV or the DISH Network.

When the bandwidth is available or affordable, less compressed or uncompressed transmission may be used. Compressed transmission using higher than consumer bit rates is "mezzanine level" compression. Uncompressed systems may use IP, asynchronous transfer mode (ATM), multiprotocol label switching (MPLS), or dynamic synchronous transfer mode (DTM) transmission protocols of synchronous optical networking (SONET) fiber-optic networks.

Figure 7.13 shows how various distribution technologies are used to create long-haul transmission paths.

A Super Headend (SHE) is an example of how multiple transmission technologies are used to aggregate content to a network operations center for delivery to consumers (see Figure 7.14).

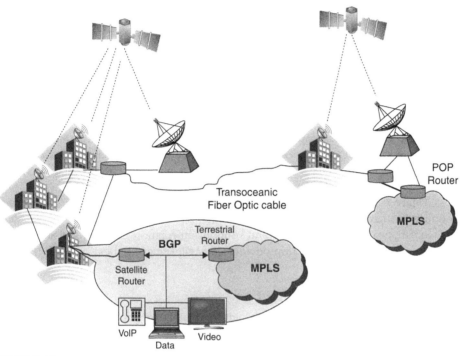

FIGURE 7.13

Satellite–cable global distribution architecture

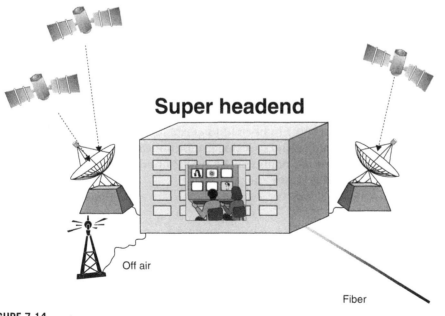

FIGURE 7.14

The Super Headend: satellite–cable distribution topology

Distribution protocols

Not all network routing protocols are suitable for TV content transmission. The selection of protocols used on a network has an impact on file transfer performance. Among the frequently used protocols are Border Gateway Protocol (BGP), MPLS, ATM, and DTM.

Border gateway protocol

The BGP (RFC 1771) is a protocol used for exchanging routing information between autonomous systems. An autonomous system is a network or group of networks that use the same routing policies and are administered as a group. Routes learned using BGP have attributes that are used to determine the best route to a destination.

In Figure 7.14, we can see that there are a number of routes that can be used to transport data from source to destination. All nodes in the network use BGP. The exchange of network metrics and routing information enables decisions to be made by network interconnection devices that will select the best path for the data transmission.

Multiprotocol label switching

MPLS is a packet-switched protocol that aids in maximizing network performance and is interoperable with IP, ATM, Ethernet, and Frame-Relay network protocols. Improved transfer speeds are attained by setting up a defined path for a sequence of packets.

The entry and exit points of an MPLS network are called label edge routers (LERs). An MPLS header containing a label is added when a packet enters the MPLS network and is removed when the packet exits. Label switch routers (LSRs) perform routing based on the label.

LSRs examine a label lookup table to determine where to switch the packets. The next hop selection operation takes place within the switched fabric and not in the router's CPU. MPLS can be considered as a combination of Layer 2, packet switching, and Layer 3, determining next hop via a lookup table for a switched path. Label Distribution Protocol (LDP) facilitates label distribution between LERs and LSRs. LSRs regularly exchange label and reachability information.

Label switch paths (LSPs) can be used to create network-based IP virtual private networks or to route traffic along specified paths through the network. LSPs are analogous to permanent virtual circuits (PVCs) in ATM or Frame-Relay networks but are independent of a particular Layer 2 technology.

Frame relay

Frame relay is a data link (Layer 2) packet-switched network protocol. It was developed as an extension of Integrated Services Digital Network (ISDN) and X.25 packet-switching technology. The driver behind its development was to find a way to transport circuit-switched protocols over packet-switched networks.

Network providers implemented frame relay for voice and data. Frame relay uses packet encapsulation over a dedicated connection. Data enters a Frame-Relay network over a connection at a Frame-Relay node.

Frame relay places data in variable-size units called "frames" and leaves error correction (such as retransmission of data) up to the end points. The structure of a basic Frame Relay frame is shown in Figure 7.15.

In 1990, Cisco Systems, StrataCom, Northern Telecom, and Digital Equipment Corporation developed a set of Frame-Relay enhancements called the local management interface (LMI). The LMI adds extensions for managing internetworks, including:

- global addressing;
- virtual circuit status messages; and
- multicasting.

Figure 7.16 shows the structure of an LMI Frame relay frame.

Frame-relay devices fall into two broad categories: data terminal equipment (DTE) and data circuit-terminating equipment (DCE). DTEs are typically located on the premises of a customer. Terminals, personal computers, routers, and bridges are examples of DTEs. DCEs are carrier-owned internetworking devices. Figure 7.17 shows the relationship of each type of device to the network topology.

Frame-Relay networks transfer data over connection paths referred to as virtual circuits (see Figure 7.18). There are two types: switched virtual circuits (SVCs), which are temporary; and PVCs.

The Data Link Connection Index (DLCI) is a number assigned to each virtual circuit and DTE device in the Frame-Relay network. Two different devices are assigned the same DLCI within Frame-Relay network—one on each side of the virtual connection.

ATM

ATM partitions data traffic into small fixed-sized cells intended to transport real-time video and audio. The International Telecommunications Union and the ATM Forum facilitated the creation of ATM standards.

Field length in bits

8	16	Variable	16	8
Flags	Address	Data	FCS	Flags

FIGURE 7.15

Structure of a Type I Frame Relay frame

Field length in bytes

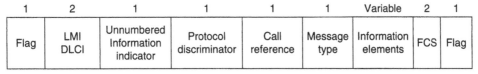

1	2	1	1	1	1	Variable	2	1
Flag	LMI DLCI	Unnumbered Information indicator	Protocol discriminator	Call reference	Message type	Information elements	FCS	Flag

FIGURE 7.16

Structure of an LMI Type II Frame Relay frame

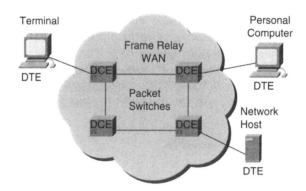

FIGURE 7.17

Typical network locations of DCE and DTC frame relay equipment

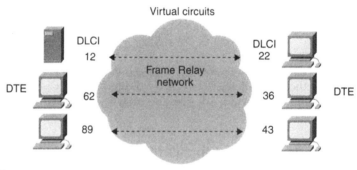

FIGURE 7.18

Frame Relay Virtual circuits

A connection-oriented technology, ATM establishes a virtual circuit between the two end points before the data transfer begins. ATM is a cell-relay, packet-switching protocol. It is a core protocol used in SONET/synchronous digital hierarchy (SDH) backbones.

ATM uses small data cells, a 48-byte payload and a 5-byte routing header, to reduce jitter (delay variance) over a network. Reduction of jitter and end-to-end delays are important for transport of real-time signals. Buffer design and size is simplified.

ATM can build virtual channels (VCs) and virtual paths (VPs). A VC uses a unique virtual channel identifier (VCI) encoded in the cell header to enable communication between two end points. A VP is a group of VCs that connect the same end points. VPs are identified by a virtual path identifier (VPI), encoded in the cell header.

ATM includes a mechanism to guarantee that "quality of service" (QoS) is maintained. QoS attributes include bit rate, packet jitter, network latency, and reliability. There are four basic types:

1. CBR: constant bit rate
2. VBR: variable bit rate
3. ABR: available bit rate
4. UBR: unspecified bit rate

Modern network management systems do not allow data rates to exceed contract limits. If this should occur, the network will either drop the cells now or flag the cell loss priority (CLP) bit for discard later in the transmission path.

DTM

DTM is a high-speed network architecture based on fast circuit switching. The packet-based protocol provides guaranteed bandwidth. The packets to be transferred are separated from other network traffic and then routed over a dedicated channel that is established between transmit and receive locations.

DTM divides fiber capacity into bandwidth intervals and can provide more bandwidth as it is needed during the transfer operation. Other circuit-switched networks, such as SONET/SDH, are unable to adjust bandwidth during a data transfer session.

Switches, rather than routers, are located at network connection points. All processing and buffering is implemented in a silicon-based switch core using application-specific integrated circuit (ASIC) devices.

Content Backhaul Over the Internet

Internet delivery of content in broadcast operations is restricted to nonreal-time applications. Segments produced by "backpack" journalists in the field and commercial spots from agencies are often delivered to broadcast operation centers via the Internet. Another suitable workflow is graphics distribution to affiliates from a centralized production location or from a production service.

SONET and SDH

As optical data transmission technology became progressively more reliable and the need arose to increase data capacity and data transfer rates, commercial carriers (AT&T in the 1970s, MCI in the 1980s) began to install fiber-optic cables in their long-haul networks. In an effort to optimize performance, protocols specific for optical transmission were developed. SONET and SDH are multiplexing protocols for transferring multiple bitstreams over the same optical fiber.

SDH was standardized (G.707 and G.708) by the International Telecommunication Union (ITU). SONET is specified in T1.105 by the American National Standards Institute. SONET is used in the United States and Canada, while SDH is used in the rest of the world.

SONET and SDH are circuit switched . Each circuit has a CBR and delay. Both are TDM, physical layer (Layer 1) protocols utilizing a permanent connection.

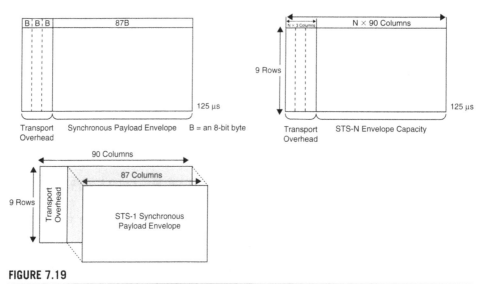

FIGURE 7.19

SONET frame structure

Table 7.2 SONET Signal Bit-Rate Capacity

Level	Bit rate (Mbps)	Multiplex capacity
STS-1, OC-1	51.840	28 DS1s or 1 DS3
STS-3, OC-3	155.520	84 DS1s or 3 DS3s
STS-12, OC-12	622.080	336 DS1s or 12 DS3s
STS-48, OC-48	2488.320	1344 DS1s or 48 DS3s
STS-192, OC-192	9953.280	5376 DS1s or 192 DS3s
STS-768, OC-768	39813.12	21504 DS1s or 768 DS3s

The basic frame format is two dimensional and illustrated in Figure 7.19. Eighty-seven payload bytes combined with three transport overhead bytes make a 90-byte row. Nine rows make a frame.

Table 7.2 lists the bit rates used by SONET. SDH has a similar hierarchical bit-rate structure but with slightly different data rates. The difference is due to the fact that each protocol was derived from the regional telephone system.

As of 2007, OC-3072 is under development and will support 159.25248 Gbps data rates.

A SONET and SDH multiplex is built by using optical combiners. The combiners can be connected in many ways. Figure 7.20 shows a number of configurations that use signal formats discussed earlier in this chapter.

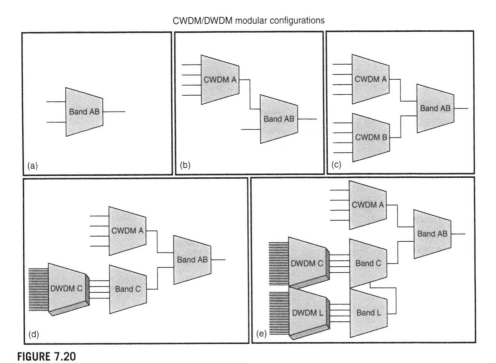

FIGURE 7.20

SONET multiplexing, CWDM/DWDM modular configurations

SONET technology can be deployed in any of the popular physical layer architectures as shown in Figure 7.21.

The ring architecture is used extensively in MANs and WANs. It offers resiliency in dual connections between add/drop multiplexers (ADMs) and has an innate backup route if the primary link fails.

SONET and SDH are multiplexes of light at various wavelengths, sometimes referred to as colors. An ADM does exactly what its name describes. A particular wavelength of light is filtered out of the multiplex; that is, it is dropped. Another signal, at the same wavelength, is added to the multiplex.

NETWORK TOPOLOGY

The network topology of the commercial carriers and the Internet can be pictured as layers, known as tiers of service. A definition of internetwork "tiers" has never been established by a standards setting body. Definition of each tier is based on financial considerations for network access.

- Tier 1: a network that can reach every other network on the Internet without purchasing IP transit or paying settlements. Primary carriers over SONET are AT&T, MCI, and Sprint.

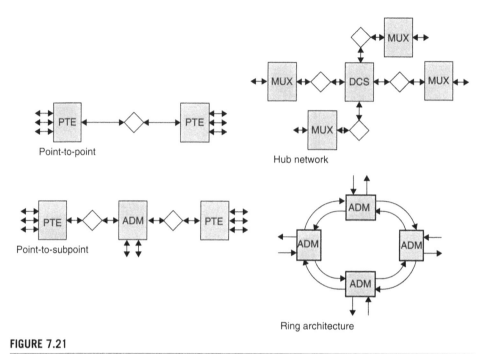

FIGURE 7.21

SONET architectures

- Tier 2: a network that peers with some networks, but still purchases IP transit or pays settlements to reach at least some portion of the Internet. Examples are Level 3, Intelsat, Eurovision, and Ascent Media.

- Tier 3: a network that solely purchases transit from other networks to reach the Internet. This tier includes typical ISPs such as EarthLink and AOL.

Geographic topology and reach

Another approach to network conceptualization is based on geography. The most localized networks are LANs. As the area covered by a network expands to a metropolitan area, the network is called, naturally, a metropolitan area network (MAN). Long-haul networks that connect MANs and cover wide geographical areas are wide area networks (WANs).

Figure 7.22 illustrates the general hierarchical topology of network categories.

Although a formal definition does not exist, LANs are generally confined to an office, building, or part of an organization. The network architecture, routing protocols, and network bandwidth are also defining characteristics. LANs were usually

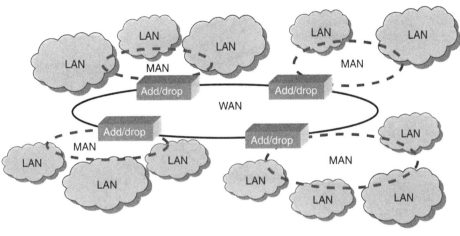

FIGURE 7.22

Network topologies (internal LAN and external WAN)

10/100 Mbps, but GigE is becoming more frequently deployed to the desktop. In the media production world, GigE is ubiquitous. A LAN can be considered an autonomous system.

MANs and WANs are optical networks. Data rates of 10 Gbps and higher are common on MANs, while WANs are now reaching 40 Gbps with 100 Gbps in the development and test stage.

In the United States, SONET-based optical networks form the backbone of many public and private networks. As can be seen in Figure 7.23, installed fibers are densest in large metropolitan areas and the MANs are hard to see. The larger lines that cross the country are the WAN backbone.

Telco TV

Telcos now are leveraging their distribution plants to deliver DTV, either over a dedicated IP broadband line or over fiber optics. Each technology used is proprietary. AT&T uses fiber-to-the-node (FTTN) for IPTV, while Verizon uses fiber-to-the-home (FTTH) and calls their technology FiOS.

Figure 7.24 shows a typical IPTV system architecture. Content for national distribution is aggregated at a national headend. The IP core network is a WAN and traverses large distances on a national scale. The IP core moves content to regional headends over a metro aggregation network, a MAN. Access nodes connect to the MAN. Content is transferred over the last mile from the access node, sometimes called a central office, to the customer. As can be seen, this distribution network architecture closely follows the WAN/MAN/LAN model.

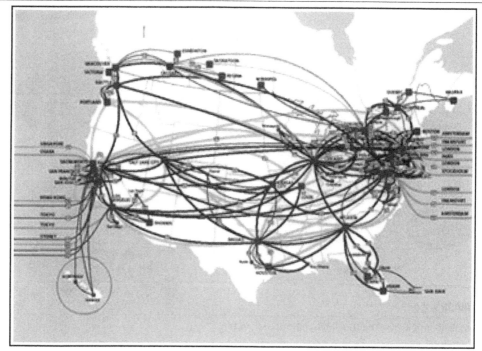

FIGURE 7.23

Typical national WAN backbone

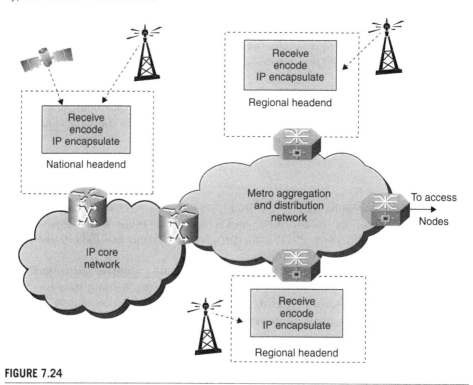

FIGURE 7.24

IPTV system architecture

Cellular backhaul

Mobile communications has been around almost as long as radio itself. Starting from 1950s, two-way radios were used in civil service vehicles such as taxicabs, police cruisers, and ambulances. The 1970s saw the consumer adoption of citizens' band radios most often used by truckers. These were all closed systems, not connected to the telephone network, but specifically suited to their purpose: reliable communication.

In time, the need to make calls from a vehicle led to mobile phones that were permanently installed in vehicles.

During the early 1940s, Motorola developed a backpacked two-way radio, the walkie talkie, and later developed a large handheld two-way radio for the U.S. military.

Mobile phones became a mainstream device over the last decade. Delivery of video either as a broadcast or as cell-based content is an evolving technological capability.

As systems grew in geographic scope, backhaul methods were needed to route calls across the country. This necessitated connecting to the public-switched telephone network: landlines.

Existing cell phone backhaul networks were originally designed for voice communications. Network topology was based on data rates of 3 kHz voice data. As features that required data capabilities emerged, improvements to the backhaul networks were made where necessary.

Consumer desire to connect with the Internet and receive video on cell phones changed everything. Even highly compressed video and audio are an order of magnitude more demanding than voice channel data capacity. This led to the development and deployment of 3G distribution to consumers and large-scale upgrades of the cellular backhaul infrastructure.

Optical backhaul over SONET networks provided a limited solution. Figure 7.25 illustrates a typical long-haul network.

Data rates of digital telephone lines were too small for the available video compression capabilities at the time and prohibited real-time delivery of TV programming. As cell phone subscribers increased, and call volume exceeded available network capacity, quality was reduced by leveraging statistical multiplexing. In addition, multiple services were aggregated over a unified transport. This topology could never support real-time video applications.

As IP-based packet transport capacities have improved, the original connection-based SONET transport networks have been supplemented by a packet transport network.

In some scenarios, the Internet can be used for nonreal-time video distribution. A typical topology is shown in Figures 7.26 and 7.27.

The three-tiered network topology is leveraged once again. Content aggregated at the national level is distributed to local operation centers. There it is supplemented with locally originating content. From there it is distributed to the tower infrastructure.

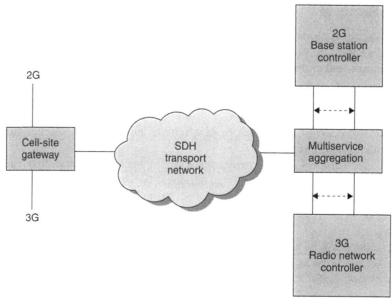

FIGURE 7.25

2G and 3G cellular backhaul optimization

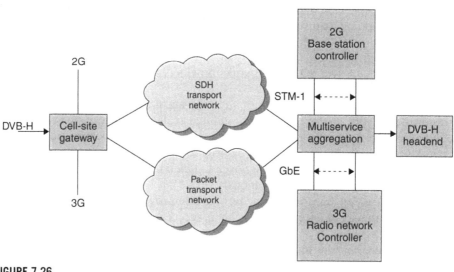

FIGURE 7.26

Cellular DVB-H backhaul integration example: SDH to packet migration scenario

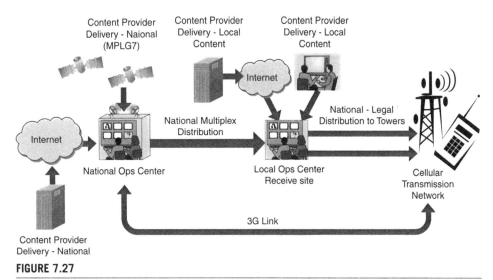

FIGURE 7.27

Cell phone content aggregation and backhaul for store and forward content distribution

THE BEST IT CAN BE

In this chapter, the movement of content between broadcast segments was described for each of the distribution networks that support a three-screen universe. The point is to emphasize the differences between professional content delivery networks and consumer distribution. As DTV screens get larger and larger, and programming uses content from many nontraditional sources, viewers will no longer tolerate low-quality presentations.

The "garbage in, garbage out" adage fits to some extent—How do you improve the quality of less-than-broadcast-quality content? At the other end of the spectrum, content produced for television employs high production values. Content distribution networks strive to deliver this content in as high a quality as possible.

8 DTV production workflow and infrastructure

Now that we've discussed how content gets into a facility and how it is distributed among facilities and consumer-facing distribution points, we can look at the infrastructure, processes, and workflows that produce content inside a broadcast operations center.

Specifically, this chapter will discuss digital production and processing for television up to the transmission phase. In terms of the content life cycle, this includes the creative and assembly phases. When content leaves assembly, it is formatted as packets of data at an appropriate data rate. This is also true for other delivery channels in the three-screen universe scenario. This chapter discusses only producing and preparing content for television.

WORKFLOW AND SYSTEM NARRATIVES

Audio, video, and graphic elements must be created, edited, ingested, stored, retrieved, archived, and assembled for broadcast. Figure 8.1 is a 50,000-foot view of a generic workflow through a production infrastructure during the creative and assembly phase.

A description of facility operations, in narrative form, is frequently developed at the beginning of the design process. It serves to ensure that all interested parties understand and agree as to what the infrastructure will do and how it will do it. A facility narrative for Figure 8.1 follows.

> *Audio and video content originates in the TV studio. ENG crews record events in the field. Content sources are "ingested" into the asset management system. This involves the storage of audio and video on server hard drives. Low-bit-rate proxies and the attachment of information needed to store and locate content—referred to as metadata—for use in library (search and retrieval) functions and editing may also be produced during this process: ingest. A media asset and workflow management application catalogs audio, video, and graphic*

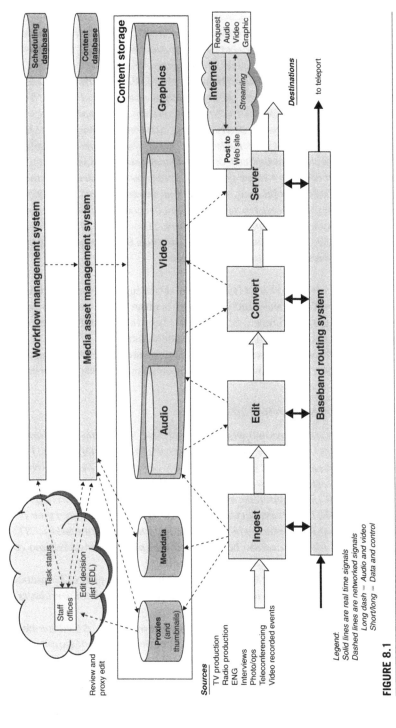

FIGURE 8.1

Content production systems and workflow

content. A workflow management application may also be implemented to keep track of scheduled ingest, editing, and other production tasks.

Experienced editors perform sophisticated "craft" edits as required to produce finished clips on editing systems. The clip, described in an edit decision list (EDL), is stored by the MAM system. In this way, the original content is never altered. Finished pieces are produced by using the EDL and writing to a medium.

The system facilitates video proxy viewing over a network. Network-available features include access to the MAM system and capabilities to locate and review content. Cuts-only proxy editing, cuts-only editing with EDL creation, and the inclusion of instructions necessary for editors to complete the task is also supported. Centralized content storage and nondestructive editing enables numerous editors to work from the same source material concurrently.

As a further aid to efficiency, a workflow management capability that includes scheduling and tracking features has been implemented. This enables the issuing of work orders for production tasks. The status of a work order as it flows through the production process can also be monitored. The clip can be reviewed and approved over the Web interface.

Content will be converted to a "house" format at ingest. Any required additional format conversions will be performed as content is moved for transmission.

A subsystem diagram (Figure 8.2) shows how the infrastructure has been portioned based on functionality to simplify detailed design.

This is the workflow overview narrative. Each system is further described in a similar way. A narrative for a baseband routing system serves as a representative example.

Baseband Routing Subsystem Narrative

The baseband routing system provides real-time distribution to audio, video, synchronization, and control signals throughout the infrastructure. The facility signal format mixes results in the implementation of four routing levels: digital video, digital audio, reference, and machine control.

Digital video will be supported in both HD and SD formats that comply with SMPTE 259 and SMPTE 292, respectively. Digital audio will conform to the AES3 standard with a sampling rate of 48 kHz. Analog format supported is NTSC composite video along with two audio channels (either stereo or dual mono) and will be converted to the house digital format at ingest.

Distribution of appropriate time code and the routing of RS422 machine control signals to VTRs, servers, and other devices comprise the other layers of the facility routing system.

Control panels will be located in TV control rooms, and at edit workstations and test stations. Functionality will consist of a combination of paging (grouping of resources by category: TEST, VTR, etc.) and x/y (assignment of an input to an output port).

All input and output signals will be available on patch panels.

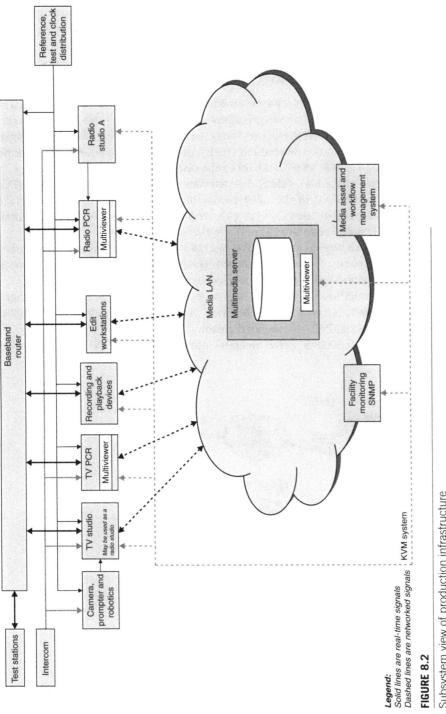

Legend:
Solid lines are real-time signals
Dashed lines are networked signals   KVM system

FIGURE 8.2

Subsystem view of production infrastructure

Now let's get a deeper take on the workflow and infrastructure used for TV content production.

THE PRODUCTION INFRASTRUCTURE

Program origination sources can be divided into three broad categories: remote, studio, and preproduced. Remote programs are backhauled to the broadcast center using techniques discussed in the previous chapter. Audio, video, and program graphics have been mixed in the production truck; commercials, logos, and other information is inserted in a MCR after backhaul to the operations center.

Studio shows mix audio, video, and graphics in a program control room (PCR). The program is then fed to the master control. The only material difference for a remote broadcast is that the PCR is not located in the operations center. Essentially, the truck at the venue and the PCR perform the same functions.

Preproduced shows do not require operator intervention. Automation systems control the playback of content, inserts, interstitials, and commercials. Figure 8.3 illustrates a simplified workflow for each scenario.

The overall workflow is considerably more complicated than the block diagram indicates. The MCR is common to all three workflows. The "deliverable" from the assembly phase is an MPEG-2 transport stream.

The next few sections delve deeper into the signal flows, workflows, and processes.

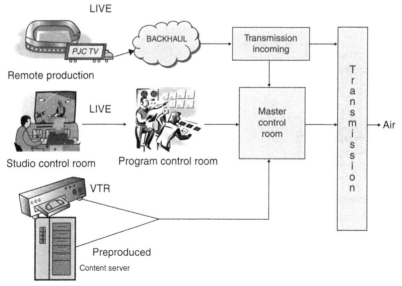

FIGURE 8.3

The three primary TV pretransmission workflows

Remote live events

Content arrives at the broadcast center from the remote site via one of the methods described in Chapter 7. If the content is compressed, it will have to be uncompressed before it can be switched in master control.

The baseband feed is routed to both an MCR and a recording device. VTRs are increasingly being replaced by server-based recording. The content will be ingested and registered in the asset management system simultaneously as it goes on air. Ingest systems also may produce low-resolution, low-bit-rate proxies that enable viewing of content over a network. Figure 8.4 provides signal-flow details.

The remote feed is routed to the master control switcher through the facility baseband routing system. After network branding and other graphics elements are added, the signal proceeds to compression. Compression codec development is an area of rapid change, but in the United States, terrestrial broadcasters are legally constrained by the FCC to use MPEG-2 video and AC-3 audio compression. Cable MSOs and satellite broadcasters are facing a related dilemma. With a large installed customer base of STB with MPEG-2 decoders, it is not possible to take advantage of new compression codecs.

Telco TV operators are not shackled by legal or installed customer base issues. With a minimal (but ever growing) number of subscribers, it makes good business

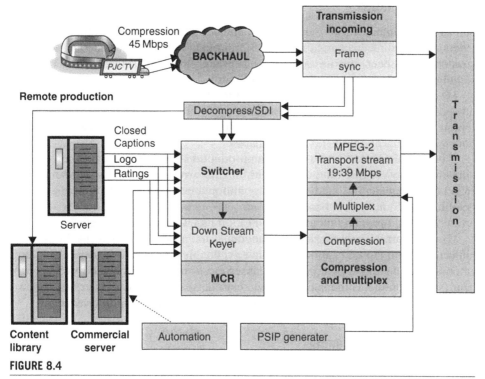

FIGURE 8.4

Remote "live" event content flow

and technical sense to use the latest compressions codecs. This is significant in that AVC and VC-1 produce comparable video quality at half the data rate of MPEG-2. This permits adding nearly double the number of services to channel offerings.

Cable and DBS operators are also installing advanced compression STBs for new customers. It remains to be seen if terrestrial broadcasters in the United States will be given the OK to move beyond MPEG-2 and AC-3 compression codecs.

At the event

A remote event can be as simple as a single camera feed from an insert studio or a single multicamera truck, or as complex as multiple trucks and the deployment of a temporary production city at the venue site.

Consider a sporting event broadcast. Talent is scattered across the venue and can be separated by a few feet or a number of miles. Cameras and microphones will be installed at each field location. A set may be built for the anchor talent.

The logistical problem is getting all the feeds back to the truck. At that point, except for the cramped quarters, it is just like any other studio show. A discussion about production of live, remote events, unfortunately, is beyond the scope of this book. A complete book is necessary to even scratch the surface.

Studio origination

Shows that originate from a studio do not have the logistical problem of getting video and audio feeds back to the truck from field crews. The studio and PCR infrastructure has been purpose driven from design, through installation and commissioning. Figure 8.5 shows the basic signal flow.

Everything that happens in a studio broadcast is controlled by a person. It is not possible to use a broadcast automation system for a live event. (This includes a remote event too.) Frame-accurate precision, as required by an automation system, is out of question. The PCR is organized chaos.

The technical heart of the PCR is the production switcher. It's a video-only device, but it will also control the audio board. In this way, audio and video sources are switched simultaneously and correctly without the need for an operator. Program graphics, VTR, and server-based clips are also mixed in the production switcher. A fully finished program mix leaves the PCR on its way to master control.

Master control room's primary function is to insert commercials, bugs, rating information, and closed captions. For a live event, this requires a master control operator. Being a mission-critical function, insertion of content is enabled over a redundant infrastructure. If the main MCR switcher fails, a downstream keyer can be used by the MCR operator to insert commercials and other important information.

Preproduced programming

Preproduced content can be delivered to transmission without any operator intervention. The entire process is controlled by a broadcast automation system.

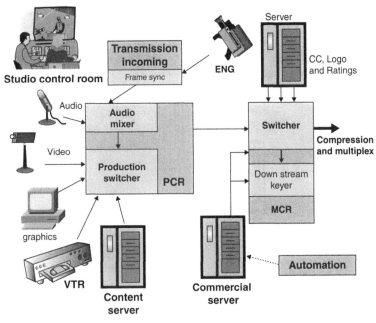

FIGURE 8.5

Studio production signal sources and flow

Figure 8.6 is a functional diagram of the signal flow. Automation is fed a rundown, also called a playlist, that has been assembled by the traffic department. It includes precise timing information for commercial insertion. The automation system is also coupled to the asset management system and can move content to playout servers or check to see if VTRs are properly cued. All the necessary information is kept in a database. A new playlist is appended to the live running list periodically. Addition functions such as ingest can be managed by the automation system.

Virtual sets

Gone are the days of wooden framed, painted canvas sets. So too are the days of 100% real sets! Today, computer systems are so powerful and application execution so fast that virtual sets (VS) are being used live, on-air with increasing frequency.

Implementing a VS system is a technically challenging undertaking. Green screen sets with minimal props in fixed locations must be tracked by the VS application along with mobile talent. Cameras, fitted with infrared sensors, constantly monitor the location of talent.

Figure 8.7 illustrates the basic visual components and assembly technique used to create a convincing virtual set.

Different virtual set providers take different approaches. Some use proprietary hardware running UNIX-based applications; others can be run only on PCs. Another approach supports systems built on open hardware, using any operating system and any GFX card.

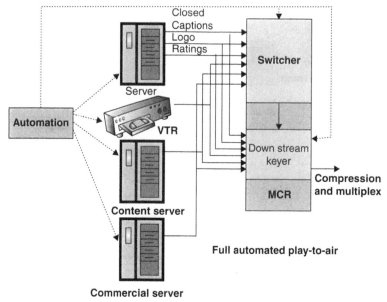

FIGURE 8.6

Preproduced media and control signal flow

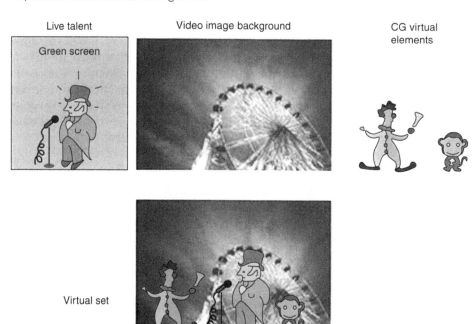

FIGURE 8.7

A scene that combines live video and virtual images

Pattern recognition algorithms recognize designated locations on studio walls and insert virtual video displays. Infrared sensors keep track of the location of talent and other on-set objects. Animation techniques such as inverse kinetics and dynamic key frames are used to create the virtual images.

As computer power and graphics and animation techniques improve, it will become increasingly difficult to determine if a scene is live or virtual.

PROGRAM ELEMENTS

TV programming consists of video and audio sources along with graphics. In the following discussion, they will be referred to as elements.

Video

HDTV, maybe the greatest invention since sliced bread, has revolutionized the broadcasting industry. Besides an increase in image detail, the most significant contribution of the effort to produce HDTV was that it accomplished the previously impossible objective of digitizing TV images in a way that enabled transmission over the existing limited bandwidth of terrestrial channels.

Presentation choices

By now terms such as 1080i, 720p, and 1080p are familiar to almost every person in the industrialized world. Even if a person doesn't know what the terms designate, they do know what HDTV is.

Presentation formats acceptable for finished HD programs ready for transmission are limited to 720p and 1080i. But those in the broadcast production profession wish that it were so simple. Rather than a choice between two formats, a plethora of production formats, all claiming to be HD, abound. I am referring to baseband signals and not to compressed video. The numerous compression formats make the situation worse.

This has placed an emphasis on format conversion. Ideally it should be avoided; no format conversion is a good conversion. Chapter 9 will discuss many of the problems involved with video format conversion. For now, it is important to be aware that format conversions during production, or for that matter, anywhere in the entire air chain, should be avoided.

Dual aspect ratio production

Even though analog television is gone, programming still may be produced using both the 16:9 HD and 4:3 SD presentation formats. This has led to the need to perform aspect ratio conversion (ARC) at many points in the production process. Beyond technology concerns, aesthetics come into play. Scenes are framed differently for a pure 16:9 aspect ratio production as compared to when content will also be converted to 4:3.

Deciding exactly when, where, and how SD and HD images should be up- or downconverted is an important infrastructure design consideration. It will have a large impact on the workflow and the distribution architecture.

Some broadcasters have answered this challenge by framing all scenes in 16:9 and producing a 4:3 SD version by letterboxing the original 16:9 image when displayed on a 4:3 screen. This results in unused areas at the top and bottom of the display.

Active format descriptor

The need to manage presentation of video in any native aspect ratio on a monitor of any aspect ratio has created a variety of formatting problems. Figure 8.8 illustrates the ideal and unacceptable scenarios.

Letterbox, pillar bars, and the dreaded postage stamp are various ways in which an image can be incorrectly displayed. It can get to the point where a postage stamp image is postage stamped a second time.

The problem was so severe that it precipitated a coordinated effort among TV standards bodies to develop a solution. The ATSC, SMPTE, and CEA pooled their technical working groups and created a method of resolving aspect ratio presentation issues.

Active format descriptor (AFD) information identifying the aspect ratio of source material is inserted in the VANC. This AFD code word is included in the user data section of the video bit stream, and persists during transmission and is decoded in the DTV receiver. Table 8.1 is the ATSC bit stream syntax, and Table 8.2 lists the association between AFD codes and receiver actions.

Although initially developed to solve consumer presentation issues, AFD information can be used during the production and assembly processes to automate ARC. Figure 8.9 is a functional diagram of a representative signal flow through an automated ARC process.

An AFD inserter tags content as it enters the facility. From this point on, ARC is performed transparently without the need for operator intervention. Before AFD, signals had to be carefully routed to and from conversion equipment by preprogrammed router controls or manual control.

Pan and scan

Automated ARC solves many image framing problems, but not all of them. The pathological case is when content of interest is at the far extremes of a 16:9 frame and will be lost in a 4:3 center cut. Figure 8.10 illustrates the problem.

Depending on the producer's intent, the visual information in the example on the right might be adequately conveyed using a split screen or multiple windows. Another technique may cut from the left to the right as dramatic action dictates.

Collaboration among ATSC, SMPTE, and CEA has developed a method to insert pan and scan instructions on a frame-by-frame basis. The information is referred to as bar data, and the ATSC syntax is listed in Figure 8.11.

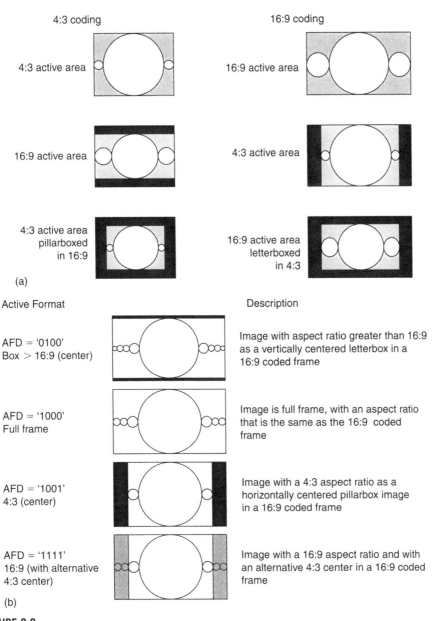

4:3 coding

4:3 active area

16:9 coding

16:9 active area

16:9 active area

4:3 active area

4:3 active area
pillarboxed
in 16:9

16:9 active area
letterboxed
in 4:3

(a)

Active Format

Description

AFD = '0100'
Box > 16:9 (center)

Image with aspect ratio greater than 16:9
as a vertically centered letterbox in a
16:9 coded frame

AFD = '1000'
Full frame

Image is full frame, with an aspect ratio
that is the same as the 16:9 coded
frame

AFD = '1001'
4:3 (center)

Image with a 4:3 aspect ratio as a
horizontally centered pillarbox image
in a 16:9 coded frame

AFD = '1111'
16:9 (with alternative
4:3 center)

Image with a 16:9 aspect ratio and with
an alternative 4:3 center in a 16:9 coded
frame

(b)

FIGURE 8.8

(a) Potential letterbox presentation issues, (b) AFD values for 16:9 coded frames (Derived from ATSC A/53 and SMPTE 2016)

Table 8.1 ATSC Bit Stream Syntax

Syntax	Number of bits	Format
user_data_start_code	32	bslbf
afd_identifier	32	bslbf
zero	1	'0'
active_format_flag	1	bslbf
reserved	6	'00 0001'
if (active_format_flag == '1') {		
Reserved	4	'1111'
Active_format	4	Bslbf
}		

Table 8.2 AFD Codes

Active_format	Description	
	4:3 coded frames	16:9 coded frames
'0000'	Undefined	Undefined
'0001'	Reserved	Reserved
'0010'–'0011'	Not recommended	Not recommended
0100	Aspect ratio greater than 16:9 (see below)	Aspect ratio greater than 16:9 (see below)
'0101'–'0111'	Reserved	Reserved
'1000'	4:3 full frame image	16:9 full frame image
'1001'	4:3 full frame image	4:3 pillarbox image
'1010'	16:9 letterbox image	16:9 full frame image
'1011'	14:9 letterbox image	14:9 pillarbox image
'1100'	Reserved	Reserved
'1101'	4:3 full frame image, alternative 14:9 center	4:3 pillarbox image, alternative 14:9 center
'1110'	16:9 letterbox image, alternative 14:9 center	16:9 full frame image, alternative 14:9 center
'1111'	16:9 letterbox image, alternative 4:3 center	16:9 full frame image, alternative 4:3 center

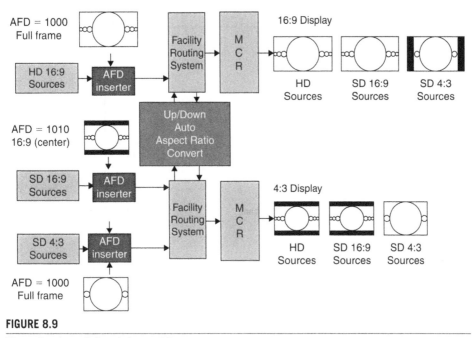

FIGURE 8.9

The active format descriptor workflow

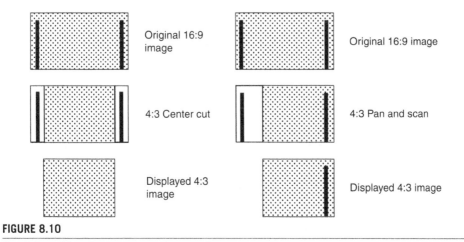

FIGURE 8.10

4 × 3 from 16 × 9 center cut and pan and scan

The technique defines the crop area in terms of marker bits. This is designated by a top line and a bottom line for the vertical dimension and left and right pixel numbers for the horizontal direction.

Syntax	No. of bits	Format
bar_data() {		
top_bar_flag	1	bslbf
bottom_bar_flag	1	bslbf
left_bar_flag	1	bslbf
right_bar_flag	1	bslbf
reserved	4	'1111'
If (top_bar_flag == '1') {		
marker_bits	2	'11'
line_number_end_of_top_bar	14	uimsbf
}		
if (bottom_bar_flag == '1') {		
marker_bits	2	'11'
line_number_start_of_bottom_bar	14	uimsbf
}		
if (left_bar_flag == '1') {		
marker_bits	2	'11'
pixel_number_end_of_left_bar	14	uimsbf
}		
if (right_bar_flag == '1') {		
marker_bits	2	'11'
pixel_number_start_of_right_bar	14	uimsbf
}		
marker_bits	8	'1111 1111'
while(nextbits() != '0000 0000 0000 0000 0000 0001') {		
additional_bar_data		
}		
}		

FIGURE 8.11

Pan scan data (From ATSC A/53)

Audio acquisition formats

Audio does not suffer from quite the same aversion to conversion as video. This is because, at least in part, audio is much less data-rate demanding when digitized.

Production audio is analog or digital. Conversion is straightforward. Analog audio is often converted to digital as it enters the routing system. The resultant PCM data can then be used to construct an AES3 pair. AES audio is a safe bet for the house audio format.

If a 5.1 surround sound mix is to be produced, six audio channels must be distributed simultaneously. This will require six cables for analog, six for PCM, and three for AES3. Dolby E processes audio so that up to eight channels can travel over an AES3-compatible signal.

Another method of moving audio through the production process is to embed it in the SDI video signal. This technique requires the installation of audio embedders and dembedders at appropriate points in the signal flow. Some routing systems offer input and output cards that can embed or dembed audio and then route it separately from the video.

Ultimately, audio will reach an AC-3 (Dolby Digital) encoder and be compressed in preparation for inclusion in the transport multiplex.

Toward an immersive experience

The HDTV visual experience is so lifelike that any unusual location cue or temporal displacement of sound will not feel or sound right; the spell of an immersive experience will be broken. Therefore, aural perception is arguably more influential in creating a life-like experience than video.

An all-too-popular technique for situation comedies, sports highlight reels, and music performances is to "sweeten" audience sounds with audio clips recorded at another instant. This is rarely very convincing. The natural audience sounds at an event have a "feel" to them. They are an integral part of those moments in space and time.

One part gray matter + two parts stimuli

Aural perception is a combination of receiving vibrations in the air at two sites (our ears) and the processing our brain does with this sensory information. Variations in level, phase, and time supply information to our brain that creates a spatial aural consciousness.

Research has been done over the decades to create a stereoscopic sonic experience where sounds heard in the listening "sweet spot" seem to originate from locations other than the two speakers. To create this sonic illusion, binaural and ambiophonic techniques are used. These techniques are based on methods that attempt to model the way the ear and brain process sound.

Divide and conquer

Surround sound acquisition can be classified into broad categories. The first division is into live and preproduced. Live events can be further broken down into studio and remote. In the studio, they can be divided into audience (game show, music event) and nonaudience (news) shows.

Depending on the content, choices need to be made about audio presentation. Should the surround sound add to the dramatic effect? Surround sound mixing is open to creative interpretation: reality, simulated reality, dramatic augmented reality, or artificial reality. Sound design and sound effects play a large role.

Let the method fit the content

At the heart of the matter is the issue of whether audio should follow video. Does it change with camera angle? Will the audience accept large fluctuations in level and quality? How will a shift from surround to stereo, especially at commercial breaks, go over? Do you fold down 5.1 to stereo with interstitial sound design? Each scenario—sports, concert, studio news, and preproduced—requires different production techniques.

For a live event, such as football, the two primary sound sources are the announcers and the crowd. Obviously, the announcer goes in the center channel. A "generative" technique can be used to position the front and surround "ambiance" mics at four corners of the stadium. With an end zone shot, do you remix the "ambiance" mics to follow the camera? Where do you mix the cheerleaders and the house PA? A cut to an official on the field will leak and result in a higher ambient crowd level. Do you use LFE to convey a "shaking" stadium? Figures 8.12 and 8.13 illustrate the two different micing perspectives.

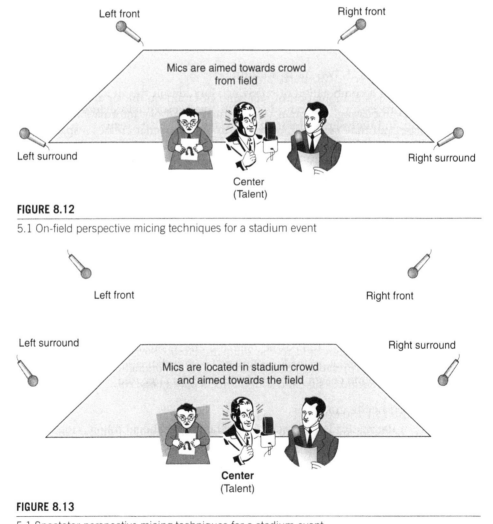

FIGURE 8.12

5.1 On-field perspective micing techniques for a stadium event

FIGURE 8.13

5.1 Spectator perspective micing techniques for a stadium event

What about music productions? Using a stereo mix as a foundation and placing lead vocals in the center (maybe with some L and R) and then adding audience ambiance to the rear speakers generally produces a pleasant effect. Should the LFE channel be used for bass sounds? Is it appropriate during a close-up of the drummer to make the drums louder? Might a creative, experimental piece explore novel use of the 5.1 channels?

Studio news—whether one, two, or three announcers—can use C, L, and R mixes to accentuate the visual spatial relationship of on-air talent. What do you mix into the rear and LFE channels? How real do I want my newscasts to be? A parade is happy; a battle is terrifying. Could public opinion be manipulated in the same way as commercial advertising influences buying habits through the use of a dramatic sonic presentation of news that would influence emotional response to perceptual stimuli?

Creating graphics

Graphics are a means of establishing the brand identity of broadcasters. Glance at a logo and you immediately recognize the network and broadcaster. Needless to say, graphics production is a core process for all broadcast operations.

Designing GFX for SD and HD

The compatibility difficulties that broadcasters face when creating graphics fall into two broad categories: presentation and color. The variables are:

Presentation

- Aspect Ratio: 16/9, HD and 4/3, SD (square pixel issue)
- Resolution: HD and SD and (analog NTSC)

Color

- Color Space: HD, SD, and NTSC
- Color Depth and Range: 8, 10, 12 bits; ATSC: 16–235; CG: 0–255 (both 8 bit)
- Color Components. Y, U, V and R, G, B

An efficient production process will create GFX once in a way that the elements can be used in both formats, and look good. This must be true regardless of whether the GFX is overlayed on an HD source, center cut, and downconverted to SD, or if it is used in an SD production and mixed in master control.

Color space conversion

Computer GFX systems represent RGB color values in various bit depths (8, 10, 12, etc.). For 8 bits, computer color space runs the range 0–255, but for ATSC DTV, the range is 16–235. Subsequently, if a GFX element is composited with video content later in the production process, nonlegal colors may be produced.

Most GFX creation applications have the capability to legalize color space to broadcast standards and store elements in RGB or convert to YUV. Color conversion issues will be discussed in the next chapter.

Postproduction systems

Once the basic audio, video, and graphics elements have been produced, they move to the next stage in the workflow: postproduction (naturally). Three key aspects of postproduction systems will be discussed: editing, compositing, and render farms.

Editing

Editing can be done using two basic techniques. One is to work with the full-resolution content. The other is work using low-resolution proxies.

The first (and older) approach downloads full-resolution video and audio to a local workstation. Once the edit is completed, the finished piece is published—stored in the main media server—and cataloged by the asset management system. This technique allows editing to be performed on full-resolution content; frame-accurate edits with elaborate transition effects are possible. When more than one editor is working with the same material, coordination of tasks becomes crucial. However, multiple version of clips must be stored and managed. Sufficient network bandwidth, at peak production times, can also become an issue.

The central portion in Figure 8.14 illustrates the process. Figure 8.14 also illustrates the proxy edit method. A discussion follows, but it is left to the reader to follow each scenario in Figure 8.14.

Proxies are low-resolution images created at ingest or by the asset management system. Their low bit rate allows them to be streamed over a network in real time. Proxy editing uses a technique where (usually) cuts-only editing is performed on low-resolution video. The editing session produces an edit decision list (EDL). The full-resolution video, kept on the main media servers, is assembled into a finished piece by applying the instructions in the EDL when the clip is copied (e.g., to a playout server). A benefit of this approach is that high-resolution content is stored in only one location and file transfers and network traffic are kept to a minimum. A drawback is not working with full-resolution content and the absence of intersegment transition effects. If a more sophisticated production is desired, the EDL and instructions can be sent to a "craft" editor for completion.

Compositing

Complicated productions that include numerous graphics, animations, and, particularly, special effects that cannot be integrated with audio and video in real time must use a different technique to create segments. This method takes all visual elements and creates a video clip frame by frame.

The technique is called compositing. There are scenarios where compositing is the preferred methodology. Since long-form programs will not be assembled on the fly (live to air), compositing is the best route to go. There is the added incentive that playout of finished clips from a server or VTR can be fully automated.

Another instance when compositing is appropriate is when a complex special effect is desired. Generation of animations and graphics and even wire frame characters is compute-intensive and cannot always be accomplished in real time. The production staff can create each element and then an editor can assemble them on

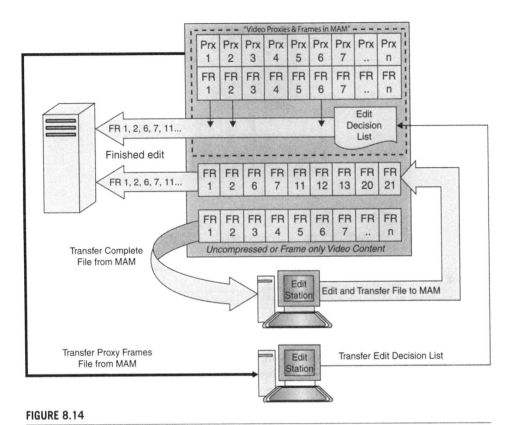

FIGURE 8.14

Local edit and publish versus proxy edit and EDL

a time line. Review of the finished product can occur before submitting the piece to the composting computers.

Render farms

Compositing, as well as format conversion and codec transcoding, is a compute-intensive operation upon which broadcasting is critically dependent. Calculating pixel by pixel, line by line, and frame by frame is compute intensive, and till the advent of multicore processor PCs, it was barely, if ever, attainable in real time.

Render farms were the precursors of multicore, parallel processing systems. All system components—the graphics card, computers (nodes), OS, applications, network, and storage—must work in harmony in a totally integrated system design for the render farm to operate at maximum capability. Shorter render times allow a larger volume of composites, animations, and more sophisticated composites and animations to be done. This gets them on the air with a minimum of production lead time.

A render farm processes video on a frame basis. The idea is to assign one video frame to each compute node for processing: in effect to use a parallel processing methodology. Figure 8.15 shows a system that is designed to convert video

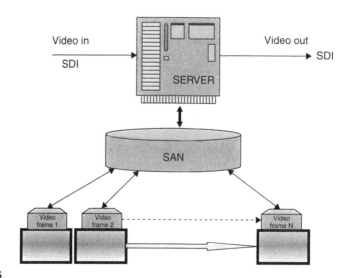

FIGURE 8.15

Render farm architecture and processing for a video presentation format conversion workflow

presentation formats. The inputs and outputs are SDI signals. For a file-based workflow, content would reside in the SAN. A composite or animation job would be queued. In a similar fashion, the render management application would assign frames to each node to process. The resultant frames are stored in the SAN.

Render farm implementation is sometimes a feature that is supported by a graphics application. If not, dedicated middleware must manage the job, parsing the file into frames and assigning them to nodes for rendering and then reassembling these new frames into a file. A centralized SAN, rather than local disks, can reduce render time dramatically; six-hour renders on a single machine can be done in less than an hour on the render farm.

When implementing a render farm, workflow must be considered. There is a trade-off between frame render time and file transfer to the farm. The requirement to do very small and short animations may be more cost-effectively accomplished with local workstation renders. But for big animations—30 seconds, full-screen HD—the less-than-real-time file transfers to the render farm are more than offset by rendering speed.

Data elements

DTV has the ability to include data elements in a program. Types of data include closed caption, ratings information, and other PSIP and PSI information tables.

Closed-caption and parental guidance rating information

Figure 8.16 is a functional diagram of an air chain that inserts data. In this representative system, logos, ratings, closed captions, and Neilsen markers are added to the baseband signal.

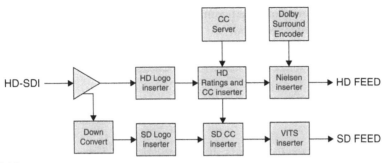

FIGURE 8.16

Precompression signal insertion

It's a fairly straightforward workflow. Insert the logo, ratings and closed captions, and Nielsen markers. Equipment vendors have produced equipment that greatly simplifies system design.

Closed-caption screen position is a serious consideration for sports in particular. Many a tavern will present multiple games on screens scattered across the establishment, all with the sound off. Frequently, closed-caption data is displayed in a way that blocks the action. Sports broadcasters need to consider the placement of CC with respect to video and graphics, or many of their prime audience members will have to endure a degraded viewing experience. Similar considerations are necessary for dramatic content.

Play-to-air content management

Broadcast TV executives are keenly aware of the key role that time slot plays in the success of a program. Careers rise and fall depending on the success of accurately identifying a winning program. Extensive research and debate goes into development and commitment to a network's programming schedule. Maybe there's a little magic and luck involved as well. Big audience numbers mean higher advertiser revenue streams.

Fortunately, the technical infrastructure that manages the scheduling and delivery of content to air is not so arcane an art as programming hit shows. The process consists of two systems: traffic and automation.

Traffic systems

The basic function of traffic management software is to build the play-to-air schedule for programming, commercials, graphics, and other content. A playlist or rundown is produced by the traffic system for import into the automation system. This is a precisely timed, frame-accurate schedule for when commercial breaks within programming will run and which advertisements will appear within them. For live programs, control room staff will manually trigger content playout. Sometimes this will consist of firing a "macro," a series of timed commands, for playout of complex sequences of audio, video, and graphic elements.

Got the time?

Broadcast timers specify time in terms of hh:mm:ss:ff—hour:minute: second:frame.

A traffic system generates reports that show what air slots are available. Ad sales uses this information to offer time slots to potential advertisers. Sales of commercial time slots are immediately shown in schedules.

Figure 8.17 is a functional flow diagram of an automation system. The process deliverable is a playlist that can be used by an automation system.

Content movement to playout devices

Traffic systems can be implemented so that content will be automatically moved to the playout server. This is accomplished by linking the frame-accurate schedules to the MAM database. Scheduling conflicts are identified and resolved. Any material that is not present in the on-air database or video server is flagged. Access to content listed in the playlist enables production personnel to quickly modify the schedule, add content, and alter timings.

One method of accomplishing automated content movement is by using the media object server (MOS) protocol. The MOS protocol has been implemented by many broadcast equipment manufacturers. MOS-enabled systems can record, copy, move, or delete content based on the playlist.

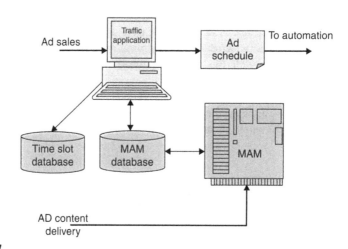

FIGURE 8.17

Elements of a generic automation system

Automation

Sophisticated combinations of audio, video, graphics, animations, and other bugs, logos, billboards, and what-have-you are sometimes beyond the ability of a technical director and their staff to take-to-air manually. In other scenarios, when programming is delivered from a server or VTR, control of all interstitial, commercials, and graphics is accomplished most efficiently without operator control, especially if multiple programs originate from the same facility.

Interstitial

An "interstitial" refers to a program of very brief duration that is slotted between two other programs of longer duration.

Figure 8.18 shows a broadcast automation system concept.

Actuation of a playout device can be achieved via a number of interfaces. VTRs are controlled over an RS422 interface. The RS422 is similar to the RS232 protocol used for nine-pin D connector interfaces on PCs but with improved EMI and RFI rejection due to the use of differential signals.

Another method of device triggering uses contact closures. In GPIs, application of a high or low voltage begins or ends playout.

Servers, being networked devices, are most easily fired over a LAN interface, generally through the control interface, in dual-ported systems.

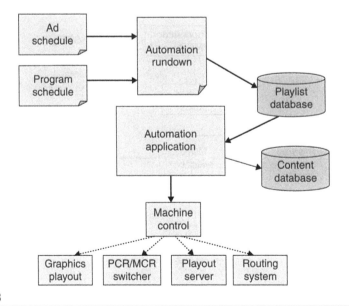

FIGURE 8.18

Playout automation system

In all cases, frame-accurate switching is required; this accounts for the need for timing based on house reference. In and out points occur in the vertical blanking interval and are synchronized by facility time code reference signal.

CREATING AN MPEG-2 TRANSPORT STREAM

Creating a DTV multiplex for transmission requires the integration of numerous processes. Efforts have been made by numerous technical bodies to automate as much of the workflow as possible. Figure 8.19 is a high-level block diagram of the systems involved in automated program play-to-air.

The tasks can be broken down into three broad areas of functionality. The first is integrating the broadcast playout automation, with traffic and media management systems. At the next stage, valid PSIP and PSI information must be produced and properly inserted into tables and transport stream packets. Finally, digital cues for downstream commercial insert must be present to automate the firing of local servers and splicers.

Different standards organizations have addressed each of these challenges. In fact, the ATS, SMPTE, and SCTE have worked together to ensure that the technologies and communication techniques are interoperable.

SMPTE's broadcast exchange format

After three years of work by SMPTE's S22-10 Working Group, SMPTE-2021, the Broadcast eXchange Format (BXF) was published in 2008. SMPTE-2021 has the

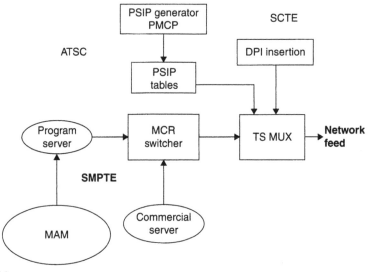

FIGURE 8.19

MCR automation process communication areas of standardization activity

potential to replace hundreds of proprietary batch and file-oriented interfaces currently in use.

BXF is a direct descendent of ATSC A/76, with many common key participants in both SMPTE and ATSC working groups. However, I'll describe it first because it is used earlier in the workflow than A/76.

Traffic, program management, and digital content distribution systems frequently use simple, sometimes manual, methods of interprocess communication. Many of these networked systems do not communicate interactively with each other, nor do they properly share metadata. BXF was developed to solve this interprocess communication problem.

The BXF standard specifies the use of XML messages for communication of three types of data exchange tasks:

1. Schedule and as-run information
2. Content metadata
3. Content movement instructions

BXF also provides:

- A single method of exchanging data among systems such as program management, traffic, automation, and content distribution
- Support for file- and message-based (dynamic) data exchange
- Increased integration of related systems
- Extended metadata set for data exchange

In addition, efforts are currently under way to develop a recommended practice for BXF deployments.

As-run monitoring

After the scheduled time for the segment has passed, the traffic system will verify which ads actually ran and generated the bills. If something went wrong, or if the advertising schedule was interrupted by an event such as breaking news, the system assists the traffic director to determine whether the ad can be rescheduled or whether the advertiser is owed a refund, credit, or make-good.

ATSC A/76: program metadata communication protocol

PSIP is the means of sending assembly instructions to DTV receivers. The information contained in PSIP tables enables frequency tuning, parsing of the MPEG-2 transport stream to find audio and video, and creation of on-screen electronic program guides. Creating PSIP tables is complicated and, if done manually, prone to errors.

The ATSC A/76 standard defines a method for communicating metadata related to PSIP table creation among systems used in the process. This includes duplicate data that needs to be entered in PAT and PMT tables required. Communication is based on XML message documents.

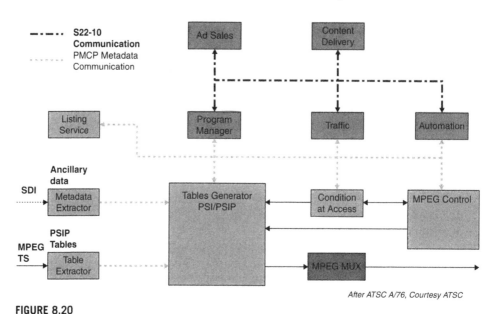

PMCP & BXF: Infrastructure Scope

After ATSC A/76, Courtesy ATSC

FIGURE 8.20

ATSC PMCP (Program Metadata and Control Protocol), and SMPTE BXF (Broadcast eXchange Format) communication scope

Communication among devices is quite complex and is illustrated in Figure 8.20.

PMCP applies to communication among table extractor, metadata extractor, and table generator, as well as among the table generator, listing service, program manager, traffic, conditional access, and MPEG control systems.

BXF is used primarily among automation, traffic, and the program manager.

Commercial insertion

The final item that needs to be inserted into the transport stream multiplex is information that enables the automated insertion of commercials as the program is distributed. In digital broadcasting, this is accomplished by the insertion of splice point information (triggers) into the transport stream. The technique is referred to as DPI and is specified in SCTE 30 and 35.

This technique enables cable headends and network affiliates to automate the process of commercial insertion by switching between two MPEG-2 transport streams, the program and the commercial. The interested reader is urged to read the SCTE standards for a detailed technical discussion of the process.

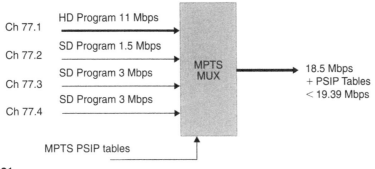

FIGURE 8.21

Typical multicast service multiplex

Tasteful commercial insertion

Programs are artistic creations. The creative team spends many hours developing, producing, and fine-tuning a property for broadcast. Of paramount importance is the insertion of commercial messages for revenue generation. Action is developed with the conscious intent to cut to a commercial at a moment of high interest or tension. Placing commercials in other temporal locations has the potential to negatively impact the effect of a program. Therefore, an issue to consider is the importance of maintaining a record of where commercials are intended to be inserted.

Multicasting

With 19.39 Mbps being the only limitation on a DTV broadcast, content delivery is not restated to a single program. Broadcasters are permitted to distribute a multicast mix of up to four programs over a single 6 MHz terrestrial channel. Figure 8.21 illustrates the technique.

This places an added burden on systems that generate and insert PSIP information. Accurate data must populate the tables for all program services; otherwise, a decoder will not be able to locate, demultiplex, and assemble associated program elements.

FROM DTV TO MULTIPLE PLATFORMS

Each of these scenarios requires thorough workflow consideration before embarking on infrastructure design. All delivery channels, all consumption devices, and all audio, video, graphics, and data must be considered in all permutations and combinations. Production workflows and the multiple-channel emission chain must be considered in light of all program elements used in any content delivery system.

Conversion processes are the means to format content for multiplatform production and diverse channel delivery. Conversion techniques are discussed in the next chapter. We move on to integrated production workflows in Chapter 10.

9

Conversion

In the last chapter we discussed typical workflows for a TV broadcast operations center. Conversions between SD and HD content were illustrated by process blocks in representative system functional diagrams.

This chapter does not purport to explain the esoteric details of digital audio and video signal processing. The mathematical foundations and implementations are the subject of intense academic and corporate research programs. Equations and algorithms are presented only in order to illustrate the complexity of conversion processing. The interested reader is urged to find expert textbooks or course offerings that will explain the topics.

TV broadcasters have grappled with converting between numerous DTV formats for nearly a decade now, and although the process continues to achieve higher audio and video quality levels, implementing conversion is still a challenge.

DTV differs significantly from analog television in three key aspects: choice of presentation formats, compression codec, and the need to include assembly instructions. DTV has abandoned the one-size-fits-all philosophy of analog television and introduced choice.

In the multidistribution channel universe, the problem is more complex by orders of magnitude. Presentation formats and compression codecs are constrained by the data rate capacity of the delivery channel, and therefore different conversion decisions must be made for each channel.

SO MANY CHOICES

First, an appropriate presentation format for audio and video must be decided upon. Even HD has two primary choices, 1080i or 720p. Audio offers 5.1 surround and a set of mixing logistics that is not of concern in stereo.

Compression may be the technology that enables all forms of digital broadcasting, but there is an ever-increasing number of codecs to choose from. In the last few years, AVC, VC-1, E-AC-3, Advanced Audio Coding (AAC), and other transmission formats have been introduced, all with increased efficiency. In the production domain, even more compression codec choices are available.

Delivery of packetized digital audio and video to any platform now requires assembly instructions. These can be implicit or explicit. What this means is that content—audio, video, graphics, text, and data—must be formatted in such a way as to be applicable to the assembly methods used by the target platform.

Conversion technology implementation

All conversion technology implementations are not created equal. The three types of system design methods (hardware, software, firmware) utilize various tradeoffs between speed and flexibility. The design compromises define the capabilities of a device to process various compression and presentation formats.

When a codec is implemented in hardware, the algorithms are hardwired into an integrated circuit. The chip is capable of executing conversion operations at full system clock speed. Hardware implementations are required for real-time processing of the higher data rate compression and presentation formats such as MPEG-2 1080p60. Besides the expense, the primary drawback is that an IC solution is totally inflexible; once the wafer is fabricated, the chip cannot be modified in any way.

At the other extreme of the flexibility scale are software-based implementations that use multicore processors or dedicated digital signal processors (DSPs). This method facilitates support for virtually any new codec by installing software upgrades. However, speed is sacrificed with a software solution; what an IC can do in a single clock cycle may take a software-based solution dozens or even hundreds of clock cycles. Real-time transcoding of data-intensive formats is impossible. However, given sufficient memory, transcoding can produce a new file in the desired codec over a period of time.

A technology that leverages both solutions uses field programmable gate arrays (FPGAs) or a similar technology to enable device speeds capable of supporting real-time processing while also offering the ability to upgrade the device via software to support new codecs.

A DTV receiver will utilize a hardware solution. This is cost effective because of the large number of units manufactured and because of the setting of legally binding standards by national regulatory agencies. In the United States the FCC DTV terrestrial transmission standard allows only MPEG-2 and AC-3 video and audio compression; a terrestrial receiver will never have to use a different codec.

PRESENTATION FORMAT CONVERSION

Audio and video presentation formats vary and present unique problems when converting between formats. Table 9.1 lists the variable attributes of an audio or video format.

Table 9.1 MPEG-2 Profiles and Levels Chart for Defined "P@L" Combinations.

	Max resolution / Max frame rate	Max bit rate	Max bit rate	Max bit rate	Max bit rate	Max bit rate	Max bit rate	
High	1920 × 1152 60fps		80 Mbps			100 Mbps	300 Mbps	1920 × 1080 60fps
High 1440	1440 × 1152 60fps		60 Mbps		60 Mbps	80 Mbps		note 2
Main	720 × 576 30fps	15 Mbps	15 Mbps	15 Mbps	15 Mbps	20 Mbps	50 Mbps	720 × 608 30fps
Low	352 × 288 30fps		4 Mbps	4 Mbps				
Max chroma sampling		4:2:0	4:2:0	4:2:0	4:2:0	4:2:2	4:2:2	
GOP supported		I, P	I, P, B	I, P, B	I, P, B	I, P, B	I, P, B	
		Simple	**Main**	**SNR scalable**	**Spatial scalable**	**High**	**4:2:2** note 1	

LEVELS — A measure of resolution and frame rate

PROFILES — A measure of decoder complexity

Note 1: 4:2:2 Profile is suitable for TV production requirements.
Note 2: 422P @ ML has increased lines / frame to include some lines from VBI.

Video formats are constructed from combinations of aspect ratio, refresh rate, pixel grid, scanning method, and color space. However, not all combinations have been implemented. Audio is a little simpler; the number of speakers and the types of services delivered vary.

Video

Video presentation format conversion is not easy. That is, high-quality video format conversion is not easy. Conversion is a melding of technology and aesthetics; some artifacts are more acceptable than others. And if you can't see a loss of quality, then it isn't there.

Video display systems have been designed based on visual capabilities. The primary consideration is the ability of the eye to resolve detail. About 1 arc minute is the accepted figure. For TV systems, this constant is used to calculate viewing distance based on screen size and pixel separation. Figure 9.1 illustrates the relationship.

On the left, there are three pairs separated by increasing distances. To the right, the dotted line represents the maximum distance at which the two pixels can be individually perceived. (Note that the drawing is not to scale.) As one would expect, when the pixels are farther apart, the optimal viewing distance moves farther away. At the bottom of the illustration, perception of the two pixels is shown for varying view distances. At the optimal distance, the two pixels touch. Get closer and they spread apart; move farther back and they blur together.

This is an important concept because video quality can only be judged when viewed at the optimal viewing distance. Too close and the image becomes a bunch of pixels; too far away and image detail disappears, a kind of low-pass filtering effect.

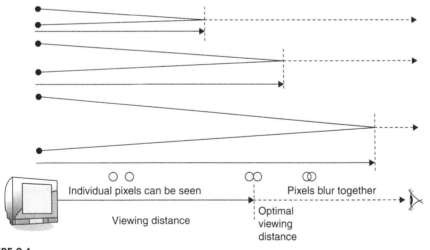

FIGURE 9.1

Viewing distance and pixel detail resolving ability

Low-pass filters

Low-pass filtering is employed to remove high-frequency noise from a digital image. The process is also referred to as "smoothing." A common application of low-pass filtering is to limit the high frequency content of an analog signal prior to digital conversion. As discussed in Appendix A, the sampling frequency determines the Nyquist limit for spectral characteristics of the analog signal.

Removal of high-frequency noise from a signal is particularly important prior to compression. The compression encoder will interpret the noise as high-frequency information and try to encode it. High-frequency content increases the bit rate in the resultant compressed bitstream. In extreme cases, this can push the encoder past its ability to encode the source content artifact-free. So, carefully designed low-pass filtering is an integral part of precompression processing. For digital signals, the filtering is done in the digital domain.

For a static image, an averaging technique can be used if several copies of an image are available. Summing the values for each pixel from each image to compute an average value will have the effect of removing random noise. However, this technique is difficult to apply to moving images.

Convolution

Convolution filters are processing tools that can be applied to image conversion. The technique uses matrix mathematics to produce a new pixel based on surrounding pixels.

$$c = a \otimes b = a \times b$$

$$c[m,n] = a[m,n] \otimes b[m,n] = \sum_{j=-\infty}^{+\infty} \sum_{k=-\infty}^{+\infty} a[j,k]b[m-j,n-k]$$

Convolution filters are a moving window operation. The operator transforms one pixel of the image at a time, changing its value by applying a function to a "local" region of pixels called "the kernel." The operator "moves" over the image to each pixel in the image.

Figure 9.2 illustrates the technique. This convolution filter applies a 3×3 kernel matrix to the neighboring pixels. The convolution function is equivalent to a matrix multiplication. A variety of results are possible depending on the kernel matrix.

Filters can be space invariant, nonspace invariant, or nonlinear. Space invariant filters apply the same operation to each pixel location. A nonspace invariant filtering changes the type of filter or the weightings used for the pixels for different parts of the image. Nonlinear, nonspace invariant filters attempt to locate edges in a noisy image before applying smoothing in order to reduce the blurring of edges due to smoothing.

Mean

Mean filters are a type of neighborhood-averaging filter. These replace the value of each pixel, $a[i,j]$, by a weighted average of neighboring pixels, i.e., a weighted sum of $a[i+p, j+q]$, with $p = -k$ to k, $q = -k$ to k for some positive k; the weights

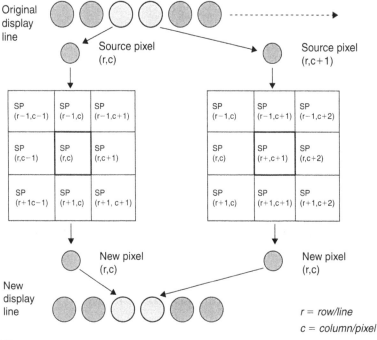

FIGURE 9.2

Pixel convolution using a 3 × 3 kernel

are non-negative with the highest weight on the $p = q = 0$ term. If all the weights are equal then this is a *mean* filter.

The simplest filter is the mean filter in which the kernel matrix entries all have a value of "1."

$$
\begin{array}{ccc}
1 & 1 & 1 \\
1 & 1 & 1 \\
1 & 1 & 1
\end{array}
$$

Convolution replaces a pixel by the average value of its neighbors. It is a simple low-pass filter that can be used for noise reduction.

Median

Median filters replace each pixel value by the median of its neighbors, i.e., the value such that 50% of the values in the neighborhood are above, and 50% are below. It can be used to remove high-frequency noise (salt and pepper). However, the removal of high frequencies will result in a loss of detail; objects which are half the size of the median filter or less will be rejected and larger objects will remain intact.

The median is calculated by first sorting all the pixel values from the surrounding neighborhood into numerical order and then replacing the pixel being considered with the middle pixel value.

For example

[8, 13, 0, 15, 27, 41, 255, 11, 39]

Produces the following pixel rank order

[0, 8, 11, 13, 15, 27, 39, 41, 255]

Sharpening

Image sharpening can be done using a Laplace filter. The technique is also referred to as edge detection.

$$
\begin{array}{ccc}
0 & 1 & 0 \\
1 & -4 & 1 \\
0 & 1 & 0
\end{array}
$$

This filter has a negative value in the convolution matrix. It sharpens the image by enhancing difference between neighboring pixels. With high-contrast images it may produce noticeable (ringing) artifacts.

Gaussian smoothing

Image smoothing can be accomplished with a Gaussian filter.

$$
\begin{array}{ccc}
1 & 2 & 1 \\
2 & 4 & 2 \\
1 & 2 & 1
\end{array}
$$

However, since pixel values are averaged, the resultant image may be blurred, depending on the filter coefficients.

Emboss filter

Many sophisticated visual effects can be produced. This convolution kernel produces an embossing effect.

$$
\begin{array}{ccc}
2 & 0 & 0 \\
0 & -1 & 0 \\
0 & 0 & -1
\end{array}
$$

Mode filters

A mode filter replaces each pixel value by its most common neighbor. This is particularly useful for *classification* procedures where each pixel corresponds to an object that must be placed into a class; in remote sensing, e.g., each class could be some type of terrain, crop type, water, etc.

Image processing

Contemporary image conversion uses a number of concatenated processes to achieve optimum image quality. Figure 9.3 illustrates a typical display processing chain.

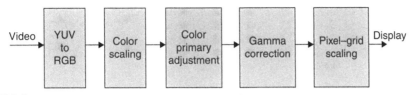

FIGURE 9.3

Image conversion processing chain

Table **9.2** Color Conversion Matrix Coefficients for Digital and Analog Television	
Digital TV	**Analog TV**
Y = 0.212 6R + 0.7152 G + 0.0722 B	Y = 0.299 R + 0.587 G + 0.114 B
R-Y = 0.7874 R + 0.7152 G + 0.0772 B	R-Y = 0.701 R − 0.587 G + 0.114 B
B-Y = −0.2126 R + 0.7152 G + 0.9278 B	B-Y = −0.299 R − 0.587 G + 0.886 B

The processes include color space conversion, color scale translation, color primary adjustment, gamma correction, and pixel-grid resizing (scaling). Display technology figures into the equation since all the parameters in the conversion process must be optimized for the target display technology. Additional processes such as sharpness enhancement and display color temperature adjustment (white balancing) add to the complexity.

Color space

Visual perception of color can be represented using a number of methods. Artists often use hue, value, and saturation (HVS) to describe color characteristics.

Color mixing is something every child learns: blue plus yellow makes green. The primary colors are red, yellow, and blue. This method of combining colors is subtractive; the color we see is the result of reflected light, the light that is not absorbed by the pigment.

Theatrical lighting and display systems add light to produce perceived colors. Red, green, and blue are the additive primary colors.

Computer graphics and TV systems don't talk the same color language. Computers process image data in RGB (red, green, blue) color space while TV broadcasts content using color represented in terms of luminance (Y) and chrominance (R-Y and B-Y), which are color difference signals. This is possible because the eye is less sensitive to differences in color than it is to differences in brightness.

Y, R-Y, and B-Y are converted to RGB and vice versa via matrix math translation. There are a number of sets of matrix coefficients used to accomplish this. Table 9.2 lists the most frequently used sets. (Note that R', G', and B' are gamma corrected values; gamma will be discussed shortly.)

Since matrix coefficient values applied at the broadcast encoder vary depending on the format of the original content, this leads to the logical query: are the correct, broadcast matrix coefficients being used during processing in the PC?

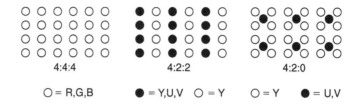

FIGURE 9.4

Color sampling formats RGB 4:4:4, MPEG-2 4:2:2, MPEG-1 4:2:0

Table 9.2 shows the difference between matrix coefficients for digital and analog television. To add to the challenge, these are not the only set of color space conversion coefficients. The inverse values are used to convert from Y, R-Y, B-Y to RGB. Get the color conversion matrix wrong and color fidelity suffers.

SMPTE EG 36-2000 details "Transformations Between Television Component Color Signals." A glance at EG 36 and the mathematical methods used for proper color format conversion reveals why there is so much confusion. Standards for color space abound. For ATSC HD, SMPTE 274/ITU 709 while ITU 601/SMPTE 259 is used for SD. NTSC color space adheres to SMPTE 170 and SMPTE 240.

Color sampling format

Another issue that can lead to loss of resolution or produce artifacts is the relationship between luminance and chrominance sampling areas. Color sampling for RGB is 1:1:1 although referred to as 4:4:4 in TV engineering jargon and there is a one-to-one correlation between RGB. Because of the fact that the eye is less sensitive to chrominance than luminance, during the conversion to Y, R-Y, B-Y, professional production systems used in compressed content workflows employ a 4:2:2 color sampling space, while MPEG relies on 4:2:0 to further reduce data for transmission. Figure 9.4 compares the techniques.

Figure 9.5 compares the spatial relationship of each sampling format in Figure 9.4. One chrominance sample in 4:2:0 color space occupies the same display area as 4 luminance pixels, hence the color detail resolution is one-fourth the resolution of luminance. It is not as bad as it appears, since the human visual system is more sensitive to luminance than chrominance.

When TV content is delivered over the Web, depending on the compression codec, a variety of color sampling spaces may be used. If you transform a 4:1:1 space using 4:2:0 methods, quality again takes a hit.

Color scaling

Another fundamental difference between DTV systems and computer graphics systems is the number and range of quantization levels for RGB or Y, R-Y, and B-Y information.

Computer systems represent RGB with a full range: for 8 bits, 0–255. ATSC DTV, due to the origins of ATV as a hybrid analog/digital technology, restricts

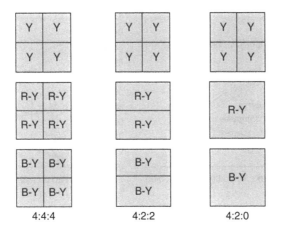

FIGURE 9.5

Comparison of color sampling techniques and spatial area

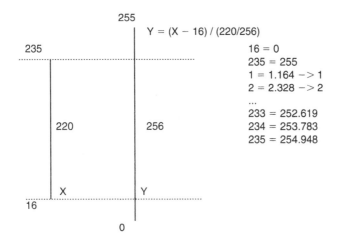

FIGURE 9.6

Comparison of ATSC DTV and PC color range

this range to 16–235 as shown in Figure 9.6. Computer graphics systems use the full 0 to 255 range. This raises havoc with the TV production chain and color "legalization" is a required step when GFX are combined with video. SDI signals use values below and above the ATSC limits for horizontal and vertical video synchronization.

ATSC range results in 219 steps as compared to 255 used by PC systems. That's a 15% reduction is color resolution. Perhaps worse is the impact of round-off errors produced when scaling TV color quantization levels to PC proportions. Cumulative errors in computational accuracy are a big problem.

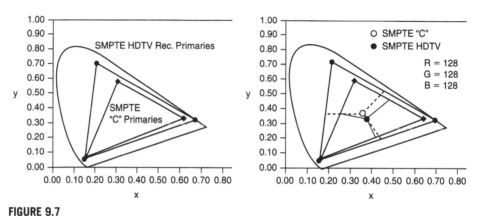

FIGURE 9.7

Color gamut. Given the same input value, different colors are produced

Color primaries and gamut

All display technologies rely on some physical means to produce light. Broadcast standards-setting bodies, such as the SMPTE and the ITU, have defined the spectral characteristics of the primary colors, red, green, and blue, that are used in imaging devices. Both ITU 601 and ITU 709 are based on CRT technology. When was the last time you saw a CRT used for a PC display?

As the graph on the left in Figure 9.7 shows, when the coordinates of the RGB color primaries are plotted on the CIE color chart, triangular color spaces are defined. As can be seen, not only will different colors be produced by differing sets of RGB values, but there are regions that can be produced with one color primary set but not with another.

Of particular concern is the point where RGB combine to form white (see the graph on the right of Figure 9.7). All whites are not identical. Whiteness is expressed in terms of color temperature: 6,500 °K is considered daylight, values below become more orange, those above, bluer. A display that is mismatched for color temperature will produce improper colors.

Gamma correction

The response to an input signal by a display, known as the transfer function, can vary. A linear relationship is ideal, however this is rarely the case. In CRTs it is exponential. This characteristic of the relationship of display output to the stimulus voltage level is referred to as the gamma. The equation (for R or any luminance of color signal) is:

$$R' = R^\gamma \quad R' = R^{2.2} \text{ (CRT)}$$

Figure 9.8 presents the gamma curves for the values of 1.0 (linear), 1.5, 2.0, and 2.5.

Gamma curves vary depending on display technology. Figure 9.8 shows a pair of complimentary gamma values that when applied at the encoder and decoder result

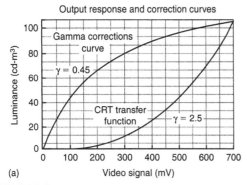

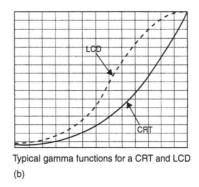

FIGURE 9.8

Gamma curves for a given value; a complimentary pair; LCD vs. CRT gammas

in an ideal gamma of 1.0. If the gamma values are not perfectly complimentary, image color will not match the source, and picture fidelity will suffer.

This presents a conversion problem for legacy NTSC content whenever any of the NTSC exclusive colors exist in an image. Color space legalizers can ensure that any video signal leaving the MCR occupies legal color space. However, there is no guarantee that the colors are aesthetically pleasing!

Pixel grids

In the three-screen universe there are many combinations to choose from.

When HD or SD broadcast content is the source display format, pixel-grid conversion is the first step in data reduction for repurposing for the Internet and handheld distribution. As shown in Table 9.3, the impact of pixel-grid conversion on data volume can be substantial. For example, a 1920 × 1080 HD pixel-grid is over 2 million pixels, while a CIF 352 × 258 display is less than 100,000 pixels. That's a reduction by a factor of 20.

Interpolation is the process of adding pixels or lines. Decimation is the process of reducing the number of pixels in a line or lines in a frame. Reducing the number of pixels in a line is also referred to as subsampling. Both processes have an impact on the ability of a format to display detail.

Pixel-grid conversion analysis

In order to analyze the effects of pixel-grid conversion, also referred to as image scaling, it's necessary to identify test patterns that will stress conversion algorithms to the max. Stimulus signals used in analog system analysis are not always best suited for digital images.

Because TV systems are described in terms of frequency, pixel-grid displays actually simplify the search for test patterns. The highest frequency that can be reproduced is when pixels or lines alternate between black and white or complimentary colors. Figure 9.9 shows the patterns used in the following discussion.

Table 9.3 Compression Codec Attribute Comparison

Codec	Video bit rate (Mbps)	Bit depth	Subsampling	Compression	Format	Standard
HDCAM	135	8	1440 Y 1080 Cb/C 3:1:1	DCT based (intra)	1080i 1080p	SMPTE 367M–368M
DVCPRO HD	100	8	1280 Y 1080 Cb/Cr	DV based (intra)	1080i 1080p	SMPTE 370M–371M
DVCPRO HD	100	8	960 Y 720 Cb/Cr	DV based	720p	SMPTE 370M–371M
HDCAM-SR	440	10	4:2:2 or 4:4:4	MPEG-4 SP (intra)	1080i 1080p 720p	SMPTE 409–2005
XDCAM@35	35	8	1440 Y 720 Cb/Cr 4:2:0	MPEG-2 (GoP)	1080i 1080p	–
DN × HD 36, 145	36,145	8	4:2:2	DCT based (intra)	1080i 1080p 720p	SMPTE VC-3
DN × HD	220	10	4:2:2	DCT based (intra)	1080i 1080p 720p	SMPTE VC-3
Infinity	50–100	10	NONE	Wavelet based (intra)	1080i 1080p 720p	JPEG2000
AVC-I	54	10	1440 Y 720 Cb/Cr 4:2:0	AVC (intra)	1080i 1080p 720p	High 10 Intra Profile
AVC-I	111	10	None	AVC (intra)	1080i 1080p 720p	High 4:2:2 Intra Profile
XDCAM HD50	50	8	None	MPEG-2 GoP L = 12 M = 3	1080i 1080p 720p	–

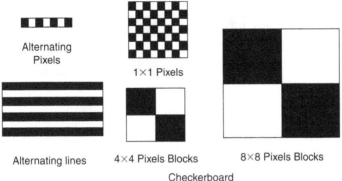

FIGURE 9.9

Alternating black and white lines, pixels, and the checkerboard pattern

FIGURE 9.10

Converting from 1440 to 720 pixels per line will have widely differing results depending on the method used

The checkerboard pattern as shown here is on a pixel basis. The concept can be employed to set conversion block size used in compression encoding. In this scenario, alternating black and white blocks of 4×4, 8×8, 16×16, and other sizes, will stress end-to-end system performance on the block level. This type of test pattern is important in analyzing where macroblocking artifacts are introduced in a processing chain.

Downconversion, pixel conversion, and aliasing

Problems encountered when using a simple technique of discarding pixels or lines as a downconversion algorithm can be illustrated by a few examples.

First consider converting a 1440-pixel line to a 720-pixel line. As Figure 9.10 shows, the result will be either an all white or all black line depending on which pixel is eliminated.

Averaging techniques fare no better and result in a gray field. In each case, all information is lost! Fortunately, real-life images rarely have this extremity of detail, but occasionally a pinstripe suit will turn to a solid color.

Upconversion

Similar issues are involved when interpolating a pixel grid from smaller to larger dimensions. The process is called upconversion. A number of techniques can be used. Regardless of the algorithm implemented it is impossible to convert an image

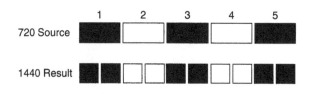

FIGURE 9.11

Conversion of a 720-pixel line to a 1440-pixel line by pixel doubling

from one resolution to a higher resolution and reconstruct all the missing detail. Some techniques are better than others, and some will use intense processing to create the illusion of improved detail, but it is impossible to turn a VGA image into a full-resolution HD image.

Pixel and line doubling

Under certain circumstances pixels and lines can be repeated to accomplish up conversion. It is not a very sophisticated technique, but is easy to implement.

Consider what happens when a 720-pixel line is converted to 1440 pixels using a pixel doubling technique.

The result of the process is shown in Figure 9.11. Whatever detail was present in the original image in a single pixel, now occupies two pixels.

One area where doubling can be used is converting 480i to 1080i pillar bar. The center portion of the 1920-pixel line occupies 1440 pixels; exactly double the 720 in an SD line. Half the display resolution is wasted, but artifacts are not produced.

Averaging

The subtle difficulties of pixel-grid conversion become apparent when doing simple calculations for various formats on a pixel-by-pixel and line-by-line basis. Ignoring the interlaced/progressive issue, only a few pixel and line ratios are simple small numbers when comparing formats.

HD to VGA

- Vertical: 1080/480 = 2.25:9 lines map to 4; or 720/480 = 1.5:3 lines map to 2
- Horizontal: 1920 center cut = 1440:1440/640 = 2:25:9 pixels map to 4
          1280 center cut = 960:960/ = 1.5:3 pixels map to 2

HD to HD

- Vertical: 1080/720 = 1.5 : three 1080 lines map to two 720 lines
- Horizontal: 1920/1280 = 1.5 : three 1920 pixels map to two 1280 pixels

DTV formats are not a perfect fit for the native resolution of PC displays. HD formats are defined as pixel grids of 1920 horizontal by 1080 vertical elements or 1280 × 720. Computer displays come in a wide variety of native resolutions. Aspect ratios of 4:3 and 16:10 are common.

Even very expensive professional graphics monitors are rarely configured for true HD pixel-grids, a 1920 × 1200 configuration often being the case. This does not

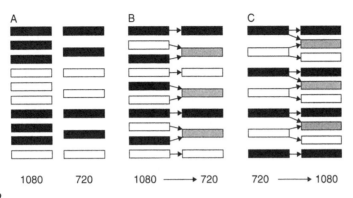

FIGURE 9.12

Conversion resolution issues between 1080-line and 720-line displays

conform to the HD ratio of 16 × 9 and a DTV image is often letterboxed with blank bars at the top and bottom of the display.

But the real problem lies in pixel-grid conversion. Even if the accumulation of errors to this point is imperceptible, the amount of damage that will be done in this during conversion is frequently a showstopper. Figure 9.12 is a visual explanation of what goes wrong.

Consider a conversion between lines that have a 3:2 relationship. As Figure 9.12a shows, when the source image consists of alternating sets of three black lines and then three white lines, the conversion is trivial and artifact-free. In Figures 9.12b and c, the image source now consists of alternating black and white lines (this is *the* killer test pattern) and conversion is no longer so simple. For every line that aligns, the next three do not. Hence, that line must be interpolated, frequently using an averaging algorithm, resulting in gray lines. Figure 9.12c also illustrates the problem when an image is upscaled: in this case, example 720 to 1080.

Here's a simple test you can try on your PC. Open any program that has drawing capabilities; Word will do just fine. Draw a box of any size and fill it with black. Now draw a line across it and change it to white. So far so good; the white line is still white. Now move the line. It turns into a shade of gray. Try it again and you get the same result. Draw another box and white lines but don't move them. Now group them and resize the box. Again the lines turn gray.

The good news is that real-world images rarely consist of such high-contrast detail at the extremes of the imaging scale. The bad news is that scaling will always produce some amount of image degradation.

As can easily be seen in the 720 to 1080 and 1440 to 720 averaging examples, more sophisticated conversion methods are necessary to preserve source image details and maintain the fidelity of the scaled image to the original scene. Who wants to watch a football game on a cell phone, played on a field without yardage lines because of poor image conversion!

Aspect ratio

DTV images can be presented in either of the two aspect ratios. In practice, aspect ratio is tied to resolution. HD resolutions are in 16×9 while SD resolutions are 4×3. However, there are other combinations. EDTV (enhanced or extended definition TV) utilizes a 16×9 aspect ratio, but with SD levels of image resolution.

In an era of global distribution and repurposing, the 14×9 aspect ratio used in Great Britain must be considered in pixel-grid conversion scenarios (Table 9.4).

Pixel squareness

The relationship of the pixel grid to the display aspect ratio determines pixel squareness. ATSC HD pixel grid dimensions have a 16:9 ratio, which is identical to the display aspect ratio. The SD pixel grid does not. This makes an attempt to map HD to SD, or vice versa, produce distorted images. Therefore a 4:3 center cut of an HD image must perform an interpolation or the resultant images will be noticeably different.

Figure 9.13 illustrates the problem, albeit in an exaggerated way. On the left, the boxes represent pixels: on the top, square; on the bottom, rectangular. The results of displaying a circle in the other format are shown on the right. From square pixels to rectangular, the circle gets fat, from rectangular to skinny, it gets fat. Images that are distorted in this fashion are said to be anamorphic.

Scanning method

Television and computers, prior to the introduction of DTV, used different display scanning methods. TV systems, in order to reduce signal bandwidth, implemented a technique that divided a video frame into fields of odd and even lines. Each full field, odd and then even, was sent sequentially. The transmission order would be:

Frame 1 Field 1, Frame 1 Field 2, Frame 2 Field 1, Frame 2 Field 2, …

The technique is illustrated in Figure 9.14.

Computer displays used a progressive scanning method that presents full frames. As will be explained in greater detail in the next section, refresh rates have a direct impact on image temporal (motion) resolution. Figure 9.14B illustrates progressive scanning.

Consider a video camera that uses sequential scanning. Once the image is captured by the camera sensor, an interlaced frame (where the interlace field rate equals the progressive frame rate) is produced by using the odd lines of the first progressive frame and then the even lines of the second progressive frame.

If line doubling is used, the resultant frames will be assembled in the following pattern:

Interlaced Frame 1 Field 1 = Progressive Frame 1, odd lines;
Interlaced Frame 1 Field 2 = Progressive Frame 2, even lines.

The big problem is the loss of spatial and temporal resolution. Line doubling was discussed earlier. In this instance the spatial detail is distributed over two fields. But there is a time difference between each field equal to half the frame rate. So not

Table 9.4 DTV Formats

System nomenclature	Luminance or RGB samples per active line (S/AL)	Active lines per frame (AL/F)	Frame rate (Hz)	Interface sampling frequency fs (MHz)	Luminance sample period per total line (S/TL)	Total lines per frame
1920 × 1080/60/P	1920	1080	60	148.5	2200	1125
1920 × 1080/59.94/P	1920	1080	60/1.001	148.5/1.001	2200	1125
1920 × 1080/50/P	1920	1080	50	148.5	2640	1125
1920 × 1080/60/I	1920	1080	30	74.25	2200	1125
1920 × 1080/59.94/I	1920	1080	30/1.001	74.25/1.001	2200	1125
1920 × 1080/50/I	1920	1080	25	74.25	2640	1125
1920 × 1080/30/P	1920	1080	30	74.25	2200	1125
1920 × 1080/29.97/P	1920	1080	30/1.001	74.25/1.001	2200	1125
1920 × 1080/25/P	1920	1080	25	74.25	2640	1125
1920 × 1080/24/P	1920	1080	24	74.25	2750	1125
1920 × 1080/23.98/P	1920	1080	24/1.001	74.25/1.001	2750	1125
1280 × 720/60	1280	720	60	74.25	1650	750
1280 × 720/59.94	1280	720	60/1.001	74.25/1.001	1650	750
1280 × 720/50	1280	720	50	74.25	1980	750
1280 × 720/30**	1280	720	30	74.25	3300	750
1280 × 720/29.97**	1280	720	30/1.001	74.25/1.001	3300	750
1280 × 720/25**	1280	720	25	74.25	3960	750
1280 × 720/24**	1280	720	24	74.25	4125	750
1280 × 720/23.98**	1280	720	24/1.001	74.25/1.001	4125	750
720 × 483/59.94	720	483	60/1.001	27.0	858	525

**Denotes analog video interface is not preferred.

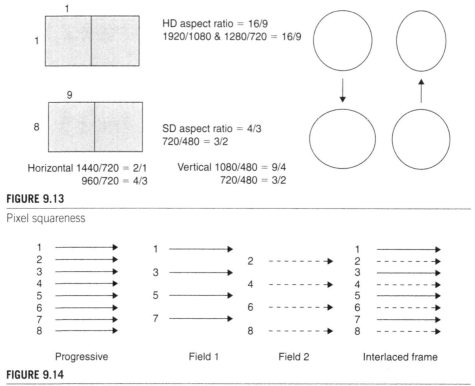

FIGURE 9.13

Pixel squareness

Progressive	Field 1	Field 2	Interlaced frame

FIGURE 9.14

Comparison of progressive (B) and interlaced scanning (A)

only is half the vertical resolution lost, but detail is lost when objects in the image are moving.

Refresh rate

This brings the discussion to the subject of refresh rate. Consider the sequence of frames in Figure 9.15. The top sequence is refreshed at 15 Hz, the middle at 30 Hz, and the bottom at 60 Hz.

Now observe how the car is moving from right to left. In the top image, the difference in the position of the car in each successive frame is farther apart than in the 30 and 60 Hz sequences. The time between each frame is 66.7 milliseconds (ms) for 15 Hz, 33.3 ms for 30 Hz, and 16.7 ms for 60 Hz. The visual system will have to fill in the blank durations or the image will appear to flicker. In fact 15 Hz is below the 24 Hz (or frames per second, fps) used by the movie industry, so this image will flicker. The other two images will appear to be continuous.

However, since the 30 Hz image is present at half the rate of the 60 Hz image, even if the resolution of the images is identical, the 60 Hz image will maintain a higher resolution for moving images; the car will not seem as blurry as it will at 30 Hz. Figure 9.16 illustrates the problem for the conversion of 60 Hz material to 50 Hz.

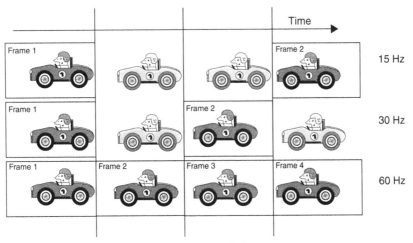

Draw frames as 16 × 9, and adjacent

FIGURE 9.15

Visual persistence and display frame rate

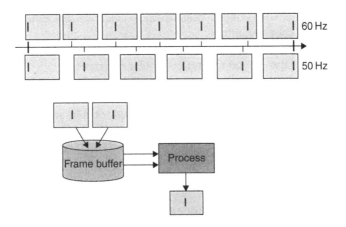

FIGURE 9.16

Converting between 60 and 50 Hz frame rates requires using techniques that maintain image feature and object persistence in the synthesized frames

The ratio of frames per second is 6 to 5. In this example, a small vertical bar moves across the display. The top sequence shows its progress for 7 frames. The sequence below is aligned such that both sequences align according to the ratio 6:5. The movement of the bar across the display is now presented for 50 Hz. It should be obvious that there is no simple way to convert from 60 Hz to 50 Hz with a simple translation function.

The frame buffer will store 6 frames of 60 Hz images and 5 frames of 50 Hz images. But, even in this trivial example, the bar will have to be extracted from the

Four film frames   Five interlaced video frames (10 fields)

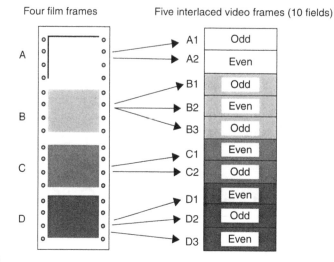

FIGURE 9.17

3:2 pulldown for film to video telecine conversion

rest of the image. This will enable an algorithm to calculate its location in the appropriate frame of the other sequence.

In the real world, images are never this basic. Now the problem becomes extremely complex; how do you extract every leaf of a tree?

Extension of the convolution filter to three dimensions is one possible approach to solving this problem. Another approach might be that if the content was compressed and then decoded, the motion vectors may enable an elegant solution to frame rate conversion. Motion vectors will be discussed in the compression section of this chapter. In any event, finding perceptually acceptable solutions to this problem have been the subject of intense research.

Film for DTV

Continuing on in this discussion of frame rate conversion, how is film to video conversion accomplished? Film is at 24 fps and video in this example is at 30 fps. Just to make things a little more interesting, video will also be interlaced.

A telecine is a machine that converts film images to video. In the digital era, it is called a datacine. The principle of operation is the same for both. Figure 9.17 illustrates the methodology.

Film frame A is used to create both fields of video frame 1. Film frame B maps to video frame 2 and to the odd field of video frame 3. The even field of video frame is created from the film frame C. Film frame C is the odd field of video frame 4. Film frame D is mapped to the even field of video frame 4 and to both fields of video frame 5. This is clearly represented in Figure 9.17.

The technique is called 3:2 pulldown. A similar method is used to create 60 Hz progressive video.

1080P24

An HD video format that is used for some transmission scenarios and for HD movies on DVD is 1080p24. This is a direct mapping of a film frame to a video frame. When decoded for display, it must undergo the same process that was just described. The format is used for two reasons. First, it is easy to convert from film to video. And, the lower frame rate produces a lower data rate when compressed. In fact, a 1080p24 video signal can fit in a 20 Mbps MPEG transport stream, with bits left over.

Word length

Increasing the number of bits used to encode a signal results in the ability to capture finer details of the source. Early digital image processing and display systems used 8 bits for each of the additive primary colors, red, green, and blue. Eight bits can quantize 256 distinct levels. These early color systems boasted the ability to discern 16 million colors. Barely adequate for photorealism, this was a great improvement over the first PCs that could display only 16 colors!

An exponential increase in data volume accompanies the increased resolution that is enabled by increasing word length. A 2-bit increase in each primary color will increase image data volume by 25%.

Data volume has an impact on the data rate produced by a compression encoder. The larger the amount of source data, the more efficient the compression engine will have to be to hit the target bit rate for the intended delivery channel.

The other concern associated with word length has to do with round-off errors accumulated during processing. Say the word length used during processing is 8 bits. After each computation the result will be rounded off to 8 bits of precision. For example, with two-digit precision, multiplying 11 times 11 will result 120, not 121. The accumulated errors can impact any conversion process. By increasing the word length during processing, only one round-off operation needs to be done, after the final calculation. This leads to better results in the processing pipeline.

Processing operation sequence

Image processing is not transitive. It is not sufficient to simply perform image decoding, reconstruction, and presentation processes randomly; the order that the processes occur in influences image quality. Processing sequences that are optimized for one display may produce artifacts on another.

Consider that for a given sequence, each processing stage degrades the image quality by 1%. This sequence of five processing steps will result in roughly a 5% quality loss; another sequence may degrade the image by 2% per stage, resulting in a 10% total loss in quality. Obviously, the first sequence is preferred.

Audio

An orphaned child no longer, audio has found a home in the DTV era. DTV receivers reproduce sound at CD quality levels. 5.1 surround systems for all tastes and budgets are available with fidelity that will satisfy every audience from audiophiles to couch potatoes.

As a starting point for discussing audio conversion, it will be helpful to make one basic calculation. The data rate for raw PCM audio in broadcast applications can be calculated by the following formula:

$$\text{Data rate} = \text{Number of channels} \times (\text{bit depth} \times 48\,\text{KHz})$$

For a single channel with 16-bit samples this results in a 768 Kbps. Increasing the bit depth, increases the bit rate proportionally: 20 bits/word = 960 Kbps, 24 bits/word = 1,152 Kbps or 1.152 Mbps.

A stereo pair will increase the data rate to around 2.9 Mbps while six full bandwidth channels are in the 6.9 Mbps ballpark.

Speaker configuration

The first type of audio conversion that will be discussed is the conversion of speaker configurations. We will consider two presentation formats: stereo or 2.0 and 5.1 surround. Figure 9.18 shows the typical arrangement of speakers for each.

Downmixing

Conversion of 5.1 surround audio to 2.0 stereo (Figure 9.19) and 1.0 mono can be achieved using relatively simple sum and difference techniques developed for stereo radio and analog TV broadcasting.

The following equations are used in the process:

5.1 to 2.0:

$$Lt = 1.0 \times L + 0.707 \times C - 0.707 \times Ls - 0.707 \times Rs$$

$$Rt = 1.0 \times R + 0.707 \times C - 0.707 \times Ls - 0.707 \times Rs$$

5.1 to 1.0:

$$M = 1.0 \times L + 2.0 \times clev \times C + 1.0 \times R + slev \times (Ls + Rs)$$

clev: center mixing level coefficient; slev: surround mixing level coefficient.

The equations offer the ability to adjust the volume relationship between the center channel and the total mix as well as the surround sound in the mix.

Coefficients could also be placed in the stereo equations to attain the same effect:

$$Lt = 1.0 \times L + clev \times 0.707 \times C - slev \times 0.707 \times (Ls - Rs)$$

$$Rt = 1.0 \times R + clev \times 0.707 \times C - slev \times 0.707 \times (Ls - Rs)$$

The 0.707 figure is the 3 dB, half-volume coefficient necessary to produce a consistent volume level when the center channel is spread over the left and right

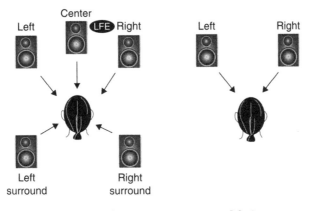

5.1 surround   **2.0 stereo**

FIGURE 9.18

5.1 and 2.0 speaker configurations

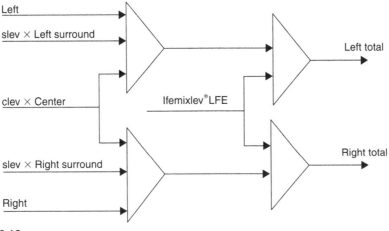

FIGURE 9.19

Surround to stereo downmix; clev and slev enable adjustment of center and surround channel levels in the stereo mix

channels. The mathematic and physics explanation is beyond the scope of this book, but the interested reader can find the information in any audio engineering textbook.

An interesting technique may be to modify the Ls and Rs balance.

$$Lt = 1.0 \times L + clev \times 0.707 \times C - slev \times (0.707 \times Ls - 0.5 \times Rs)$$

$$Rt = 1.0 \times R + clev \times 0.707 \times C - slev \times (0.707 \times Rs - 0.5 \times Ls)$$

Sophisticated, modern downmixing techniques also process the surround sound signals and add delay and reverberation to produce a spatial illusion.

Similar to dual aspect ratio video production, 5.1 and stereo conversion presents technical and artistic challenges. Should the consumer audio decoder be relied upon for the downmix? Fold down (downmix) monitoring is essential before it leaves the broadcast facility.

Upmixing

Creating a believable 5.1 mix from a stereo or mono source is significantly more difficult than downmixing. Audio processors that can perform an upconversion are available, but the question is how convincing is the 5.1 mix they can create from a stereo or mono audio source.

To get an idea of the difficulty, consider conversion of a stereo recording of a live musical event to 5.1. Analysis of the total sound content, to produce even a rudimentary upconversion, would have to extract the vocals and the crowd separately from the music. This would enable the vocals to be positioned predominately in the center speaker and the crowd in the surround speakers. A reverse vocal zapper could pull the vocals. Ambient crowd sound is noise-like and can be extracted without great difficulty.

A sophisticated vocal mix will split the vocals into lead and backing tracks, maybe a centered lead vocal with L and R backing. This will require increased extraction accuracy and a lot more processing and spatial balancing. The same issues are present for the instrumental tracks. Drums need to be spread around the mix, and the bass needs to be extracted and directed to the subwoofer. How do you create a glissando that moves from left to right across the sound field?

There are systems that process stereo audio to produce 5.1 based on psychoacoustic models of hearing. The difficulty is in making the mix sound convincingly real for all types of audio sources.

Audio services

Packetized television audio enables elements of the audio program to be delivered individually and assembled at the receiver. Each service occupies an audio channel so the more the services (see Figure 9.7) present, the higher the data rate.

If an audio program is intelligently partitioned into services, repurposing for different scenarios can be simplified. For DTV the obvious example is delivering an ME: music and effects 5.1 "service" with the voice over (VO) as a separate service. This clean feed or mix-minus can be dubbed for distribution in any language by adding a new audio service. This is the method used to create multilingual DVDs.

Tables 9.5–9.10 summarize the variable attributes of audio presentation formats.

Audio word length

Word lengths (bit depths) for audio data are longer than for video. 16, 18, 20, and 24 bit formats are regularly used. Similar to video, an increase of word length from

Table 9.5 Sample Rates (From ATSC A/52)

fscod	Sampling rate (kHz)
'00'	48
'01'	44.1
'10'	32
'11'	Reserved

Table 9.6 E-AC-3 Reduced Sampling Rates (From ATSC A/52)

fscod2	Sampling rate (kHz)
'00'	24
'01'	22.05
'10'	16
'11'	Reserved

Table 9.7 Audio Services (From ATSC A/52)

bsmod	acmod	Type of service
'000'	Any	Main audio service: complete main (CM)
'001'	Any	Main audio service: music and effects (ME)
'010'	Any	Associated service: visually impaired (VI)
'011'	Any	Associated service: hearing impaired (HI)
'100'	Any	Associated service: dialogue (D)
'101'	Any	Associated service: commentary (C)
'110'	Any	Associated service: emergency (E)
'111'	'001'	Associated service: voice over (VO)
'111'	'010'–'111'	Main audio service: karaoke

16 to 24 produces a 50% increase in data. Therefore, word length is an important consideration when processing audio for distribution on a channel with limited bandwidth, such as a cell phone or PDA.

Format conversion is only half the story but an important part. The lower the volume of audio and video data the easier it will be for a compression codec to fit the content into the delivery channel. But data volume and reduction are only part

Table 9.8 Audio Channel Configurations (From ATSC A/52)

acmod	Audio coding mode	nfchans	Channel array ordering
'000'	1 + 1	2	Ch1, Ch2
'001'	1/0	1	C
'010'	2/0	2	L, R
'011	3/0	3	L, C, R
'100'	2/1	3	L, R, S
'101'	3/1	4	L, C, R, S
'110'	2/2	4	L, R, SL, SR
'111'	3/2	5	L, C, R, SL, SR

Table 9.9 Center Mix Level (From ATSC A/52)

cmixlev	clev
'00'	0.707 (−3.0 dB)
'01'	0.595 (−4.5 dB)
'10'	0.500 (−6.0 dB)
'11'	Reserved

Table 9.10 Surround Mix Level (From ATSC A/52)

surmixlev	slev
'00'	0.707 (−3 dB)
'01'	0.500 (−6 dB)
'10'	0
'11'	Reserved

of the problem. Just as there are numerous presentation formats, there are also many compression codecs to choose from. And to add to the challenge, each codec can produce a variety of bit rates.

COMPRESSION CODECS

Even with the availability of gigabit networks, HDTV workflows using uncompressed video formats are beyond the reach of contemporary networks.

Transmission of digital audio and video is a numbers game, in that the data capacity of the delivery channel determines the characteristics of the content. For real-time content delivery, the level of image detail, the audio quality, and the ability to deliver additional data services are all dependent on delivery channel data capacity.

In nonreal-time applications, such as audio, video and document download, channel capacity translates to time. What good is being able to download an HD movie if it takes 20 hours to get it onto your hard drive? Worse still: Who has room for media files that are hundreds of gigabytes in size?

Hence, reduction of the volume of data has been an enabling technology in the digital media revolution. Reduction of media data, compression, is based on human aural and visual perception used in tandem with IT data reduction techniques.

Compression terminology

The terminology used to describe compression can get confusing. Many terms that have specific technical meanings are used loosely by technologists and management.

Compression techniques fall into two broad categories: lossless and lossy. Run length coding (RLC) and Huffman/variable length coding (VLC) are lossless: data out is identical to data in. On the other hand, quantization and coefficient weighting discard data during the compression process that can never be recovered. These techniques succeed because they are based on human perception: What cannot be perceived can be discarded.

Lossy, also known as perceptual, compression codecs can have a drastic effect on content quality if used incorrectly in a workflow. Because information is deleted with each compression/decompression cycle, concatenation of compression cycles will lead to accumulated errors that sooner or later will become noticeable and annoying.

But, I digress; back to terminology. Transfers of media in digital form have an associated data rate, specified in bits per second (bps). *Transrating* is a conversion process that changes the bit rate but uses the same compression codec. For example, an SD program that is produced as MPEG-2 at 40 Mbps is transrated to MPEG-2 at 25 Mbps for distribution to affiliates and then transrated again to 4 Mbps for multiplexing into an MPEG TS for DTV broadcast.

Transcoding is the converting of content from one compression methodology to another. MPEG-2 to MPEG-4 and AC-3 to AAC are examples of transcoding processes. Although MPEG-2 compression is required for ATSC terrestrial DTV broadcasts; AVC, VC-1, AC-3, MP3, and other recently standardized codecs are available for use in various stages of the content production and distribution chain.

It is important to clearly understand the meaning of each of these terms and to apply them properly and consistently. For example, some technologists will describe the process of converting HD to SD as a downconversion. Others will use the term "downconversion" to describe a bit rate reduction, which should properly be called transrating.

Compression standards offer a tool kit of techniques that may or may not be in the content compression process. The implementation philosophy was to require as little complexity as possible in the decoder and put all the heavy compression processing in the encoder.

MPEG-2 classifies formats as combinations of profiles and levels. Profiles characterize the presentation characteristics of the content. They are specified as maximums. For example, MPEG-2 "Main" level tops out at 720 × 576 at 30 fps, enough to support 480i images. "High" level is intended for HD and sets the maximum display and refresh rate at 1920 × 1152 at 60 fps, 1080p60. Table 9.1 showed the various combinations; note that some combinations are not allowed and denoted by empty cells.

Profiles specify color sampling and compression frame types. "Main" profile uses 4:2:0 and I, P, and B compressed frames. The "4:2:2" profile also uses I, P, and B frames, but 4:2:2 color sampling; it is intended for professional production.

Combinations of profile and level define a codec implementation and maximum bit rate. For example, SD is main profile at high level "MP@HL" with a maximum bit rate of 15 Mbps. HD is MP@HL with a maximum bit rate of 80 Mbps; both use I, P, and B frames and 4:2:0 color sampling. By contrast, 4:2:2 at high level tops out at 300 Mbps.

Quality vs. quantity

The decoded quality of video and audio produced by an encoding or conversion device is dependent on the number of bits that are produced per second. Scene complexity (more detail and more motion) and sonic complexity (more sound, random tonality) and noise take more bits to compress. Using too few bits may produce perceptible artifacts. The relationship between perception and image quality is complex and subjective.

For MPEG video encoders, at least in theory, as many bits as necessary are used to compress a given frame as required, because each scene varies in spatial complexity and the difference between frames varies; hence, there is a variable number of bits in a compressed frame. On the other hand, audio is often coded using a constant number of bits over a defined time period.

Variable bit rate (VBR) encoders produce bursts of data at changing rates. Data produced is scene dependent. A complex scene needs a maximum number of bits, while simple scenes require significantly less. Buffer design is complicated because data quantities vary over time. A target buffer in the decoder places a constraint on an encoding device and manages buffer use. This use of VBR encoding produces the highest quality content when decoded. There is no direct way to calculate a frame boundary.

The ATSC implementation of the MPEG TS specification constrains the data rate to a fixed 19.39 Mbps, a constant bit rate, yet there is confusion as to the use of the phrase. The DVB-T specification limits the TS to a maximum data rate of 24.1 Mbps.

Constant bit rate (CBR) is when an encoder produces a constant data rate bitstream. This does not mean each video frame has a constant number of bits. There is no way to know where a frame boundary occurs in the bitstream by using a simple mathematical relationship. The bitstream must be parsed and analyzed to determine frame boundaries. One of the drawbacks of CBR is that complex content that requires bit rates exceeding the limit rate will suffer quality loss. On the positive side, CBR simplifies buffer design.

Another technique tries to leverage characteristics of both VBR and CBR. Capped variable bit rate (CVBR) uses VBR encoding whenever possible, but when scene complexity requires more bits than are available, a maximum limit caps the bit rate. This may produce artifacts, yet the probability of the need for more bits than are available is relatively low. CVBR can be used in multicast scenarios to insure that an HD program attains full HD quality the majority of time, and does not impact simulcast SD programs. Additionally, during scenes of low complexity, opportunistic data can be delivered for enhanced DTV services.

Display format bit rates

Table 9.11 presents the various display formats and their associated data rates. It is important to note the number of bits per frame in order to understand how widely data rates vary. The difference in the number of pixels from SD to HD display resolutions is an increase of three to six times, and, as the table illustrates, frame refresh rate (along with scan method) can impact bit rate.

If a delivery channel can sustain 10 Mbps, the only uncompressed format that can be reliably transmitted is QCIF at a refresh rate of 15 Hz; its bit rate is 6 Mbps. If a compression encoder has a coding gain of 10:1, more format options are available. QCIF at 15 Hz will have a bit rate of 600 Kbps. The highest quality images that can be delivered over the 10 Mbps channel are VGA at 15 Hz, or CIF at 8.7 Mbps. If one can live with motion judder, SD can be delivered at 8.4 Mbps. Now increase the channel capacity to 20 Mbps and the compression gain to 50:1. This data rate is high enough to support both 1080i and 720p HD presentation formats.

Table 9.11 Bit Rates for Various Pixel Grids and Refresh Rates							
Format	Pixel grid	Pixels	YUV B/frame	Bits/frame	15 Hz Mbps	30 Hz Mbps	60 Hz Mbps
HD	1920 × 1080	2,073,600	4,147,200	33,177,600	500	995 (1080i)	1991 (1080p)
HD	1280 × 720	921,600	1,843,200	14,745,600	221	443	885 (720p)
SD	720 × 480	345,600	691,200	5,529,600	84	166 (480i)	332 (480p)
VGA	640 × 480	307,200	614,400	4,915,200	74	150	300
CIF	352 × 258	90,816	181,632	1,453,056	22	44	87
QVGA	320 × 240	76,800	153,600	1,228,800	18	37	74
QCIF	176 × 144	25,344	50,688	405,504	6	12	24

Compression for production

As stunning as the images and sound produced by high-definition DTV systems may be, it is a technical challenge to produce sophisticated segments and programs using tape-based technology and workflows. Tape is a linear medium. This applies to two aspects of the production workflow. Content on a tape must be accessed in a linear fashion; if the desired clip is at the beginning of the tape, it can be found quickly; if it is at the end, even using fast forwarding, it still may take a while to locate the desired scenes. Once the content has been found, using tape-based editing, only one editor can work on a clip at a time. When they are done, they physically, via "sneaker net," hand it off to the next person who will work on the piece.

Compressing content and storing it as a file has many advantages over tape-based production. Specifically, content can be accessed in a random manner and more than one editor can work with it at the same time. The technique, nonlinear editing, has become the production norm. Advent of file-based acquisition now extends back to the camera.

This poses a difficult set of requirements for any compression codec to meet if it is to be suitable for production.

The key to encoding source material at a high-enough bit rate so that the errors accumulated from compression cycles as the clip moves through postproduction are not visible. However, the bit rate must be low enough so that the content can travel across the production network in an acceptable amount of time. It must be orders of magnitude faster in order to make file-based production cost effective. The good news is that there are production codecs that meet both requirements.

As can be seen in Table 9.12, the bit rate choices range from 36 Mbps to 440 Mbps. The higher the bit rate, the longer the video quality will remain above the perceptible artifact threshold through successive compression cycles.

MPEG-2 video compression

As we move into a discussion of video compression, it is assumed that the reader is familiar with the MPEG compression process and terminology. For those that need to get up to speed, a Web search for an "MPEG" or "video compression" tutorial will return a wide variety of documents and Web sites on many levels of technical details.

Table 9.12 Production Compression Formats

Format	Avid DN × HD 36	Avid DN × HD 145	Avid DN × HD 220	DVCPRO HD	HDCAM	HDCAM SR
Bit depth	8-bit	8-bit	8- and 10-bit	8-bit	8-bit	8-bit
Sampling	4:2:2	4:2:2	4:2:2	1280 Y samples 4:2:0	1440 Y samples 4:2:2	4:2:2 or 4:4:4
Bandwidth	36 Mbps	145 Mbps	220 Mbps	100 Mbps	135 Mbps	440 Mbps

Figure 9.20 is a functional diagram of an MPEG compression encoder. The well-known I frame processing steps are DCT transform, quantization, Huffman coding, and RLC. P and B frames are produced using a feedback mechanism that results in the encoding of the differences between frames. Motion estimation produces motion vectors that further aid in data rate reduction.

The sequence of frames produced beginning with an I frame and continuing until the next I frame is called a group of pictures or GOP. A long GOP is commonly used in DTV broadcast transmissions and can be represented as:

I B B P B B P B B P B B P B B

This sequence of 15 compressed frames is called a long GOP.

Table 9.1 details the various profiles and levels along with the maximum compressed bit rates for various combinations of profiles and levels.

Next-generation video codecs

MPEG-2 was designed for DTV, but actually it was adapted for HDTV. During the development of MPEG-2 in the early 1990s, the Internet was not a means of mass consumer communication. But it was hoped that broadcast quality DTV, compressed form, would eventually be delivered over the Internet. It has been a long wait. But reaching the goal is at hand.

The latest generation of advanced codecs contains combinations of profiles and levels that are appropriate for the Internet, cell phones, and mobile DTV as well as full bandwidth terrestrial, satellite, cable, or telco fiber delivery of television.

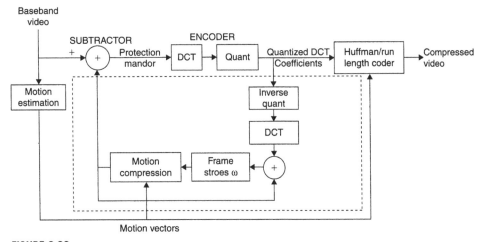

FIGURE 9.20

Motion compensated MPEG compression engine

Two advanced compression technologies have been standardized and are finding application in distribution of TV over traditional and new media delivery channels.

H.264/AVC/MPEG-4 Part 10

Developed as joint effort between ITU and ISO, MPEG-4 Part 10, also known as H.264 or AVC, achieves video quality equal to MPEG-2 at about half the bit rate. But with the inclusion of levels appropriate for Internet and mobile delivery, the codec can be used as a one-stop shop for compressing video for three-screen distribution.

One feature that is a noteworthy improvement over MPEG-2 is the ability for the encoder to select a lossless data reduction algorithm based on scene content. MPEG-2 used only one technique.

A detailed discussion of AVC/H.264 is beyond the scope of this book. However Table 9.13 illustrates the versatility of the codec. An expanded range of levels offers pixel grid sizes that are appropriate for each of the three screens. The seven profiles support not only consumer delivery channel bit rates, but professional color space and the high bit rates necessary for top quality production.

VC-1

In an unprecedented gesture that amazed the professional broadcast industry, Microsoft submitted its Windows Media 9 Video codec to the SMPTE for standardization. Prior to this event, Microsoft carefully guarded the algorithms used in its media players. The result is the SMPTE standard, VC-1. Although high bit rates are available, it has been used primarily in consumer applications.

As Table 9.14 shows, all profiles use the 4:2:0 color sampling format and an 8-bit word length. However, unique to VC-1 is a processing tool specifically designed to handle scene transitions. These are difficult for an MPEG codec to compress efficiently because of the large amount of bits required when there are large changes in an image from frame to frame.

Codec comparison

AVC and VC-1 are more sophisticated than MPEG-2. By the numbers, in the real world, their performance cuts bit rates in half for equal image quality. Table 9.15 compares the three codecs feature by feature.

Compared to MPEG-2, in almost every category, AVC and VC-1 offer advanced tools. But with this power comes complexity. AVC and VC-1 encoding is a complicated process. There are many choices that must be made in codec design. This can lead to widely varying quality in codec implementation.

But from the perspective of a three-screen universe, the fact that these advanced codecs can produce compressed bitstreams for transmission over each channel is a huge advantage over MPEG-2.

Table 9.13 MPEG-4 Part 10 AVC (H.264/AVC) Profiles and Levels Chart for all "P@L" Combinations

LEVELS	Max resolution / Max frame rate	Max bit rate	Max bit rate	Max bit rate	Max bit rate	Max bit rate	Max bit rate	Max bit rate
5.1	2k × 1k/4k × 2k 120fps/30fps	240 Mbps	240 Mbps	240 Mbps	300 Mbps	720 Mbps	960 Mbps	960 Mbps
5	2k × 1k 72fps	135 Mbps	135 Mbps	135 Mbps	168.75 Mbps	405 Mbps	540 Mbps	540 Mbps
4.2	1920 × 1080p 60p (60fps)	50 Mbps	50 Mbps	50 Mbps	62.5 Mbps	150 Mbps	200 Mbps	200 Mbps
4.1	HD 720p/1080i 60p/30i	50 Mbps	50 Mbps	50 Mbps	62.5 Mbps	150 Mbps	200 Mbps	200 Mbps
4	HD 720p/1080i HD 60p/30i	20 Mbps	20 Mbps	20 Mbps	25 Mbps	60 Mbps	80 Mbps	80 Mbps
3.2	1280 × 720p 60p (60fps)	20 Mbps	20 Mbps	20 Mbps	25 Mbps	60 Mbps	80 Mbps	80 Mbps
3.1	1280 × 720p 30p (30fps)	14 Mbps	14 Mbps	14 Mbps	17.25 Mbps	42 Mbps	56 Mbps	56 Mbps
3	SD 525/625 30i/25p	10 Mbps	10 Mbps	10 Mbps	12.5 Mbps	30 Mbps	40 Mbps	40 Mbps
2.2	SD 525/625 15fps	4 Mbps	4 Mbps	4 Mbps	5 Mbps	12 Mbps	16 Mbps	16 Mbps
2.1	HHR 480i/576i 30i/25p	4 Mbps	4 Mbps	4 Mbps	5 Mbps	12 Mbps	16 Mbps	16 Mbps

A measure of resolution and frame rate

(Continued)

Table 9.13 (Continued)

A measure of resolution and frame rate — LEVELS

	Max resolution / Max frame rate	Max bit rate	Max bit rate	Max bit rate	Max bit rate	Max bit rate	Max bit rate	Max bit rate
2	CIF 352 × 288 30 fps	2 Mbps	2 Mbps	2 Mbps	2.5 Mbps	6 Mbps	8 Mbps	8 Mbps
1.3	CIF 352 × 288 30 fps	768 kbps	768 kbps	768 kbps	960 kbps	2.3 Mbps	3.07 Mbps	3.07 Mbps
1.2	CIF 352 × 288 15 fps	384 kbps	384 kbps	384 kbps	480 kbps	1.15 Mbps	1.54 Mbps	1.54 Mbps
1.1	CIF/QCIF 7.5 fps/30 fps	192 kbps	192 kbps	192 kbps	240 kbps	576 kbps	768 kbps	768 kbps
1b	QCIF 176 × 144 15 fps	128 kbps	128 kbps	128 kbps	160 kbps	384 kbps	512 kbps	512 kbps
1	QCIF 176 × 144 15 fps	64 kbps	64 kbps	64 kbps	80 kbps	192 kbps	256 kbps	256 kbps
	Max chroma sampling	4:2:0	4:2:0	4:2:0	4:2:0	4:2:0	4:2:2	4:4:4
	GOP supported	I, P	I, P, B	I, P, B, SI, SP	I, P, B, SI, SP	I, P, B, SI, SP	I, P, B, SI, SP	I, P, B, SI, SP
	Max sample bit depth	8 bits	8 bits	8 bits	8 bits	10 bits	10 bits	12 bits
		Baseline	Main	Extended	High (HP)	High 10 (Hi10P)	High 4:2:2 (Hi422P)	High 4:4:4 (Hi444P)
						FRExt Profiles (Fidelity Range Extensions)		
		PROFILES						

A measure of decoder complexity

Table 9.14 VC-1 Profiles

Profile	Level	Chroma format	Max. sample depth	Max. bit rate (bps)	Interlace support
VC-1 profiles					
Simple	Low	4:2:0	8 bits	96 K	
	Medium	4:2:0	8 bits	384 K	
Main	Low	4:2:0	8 bits	2 M	
	Medium	4:2:0	8 bits	20 M	
	High	4:2:0	8 bits	20 M	
Advanced	L0	4:2:0	8 bits	2 M	
	L1	4:2:0	8 bits	10 M	X
	L2	4:2:0	8 bits	20 M	X
	L3	4:2:0	8 bits	45 M	X
	L4	4:2:0	8 bits	135 M	X

Compressed domain processing

Processing compressed content can be done in two ways. The brute force method completely decodes the content to baseband, performs whatever processing necessary to achieve the desired result, and then compresses the content (see Figure 9.21). Since perceptual compression is based on discarding information that cannot be perceived, each decompress/compress cycle discards information. The cumulative effect can be noticeable artifacts.

A more elegant approach, now possible because of increased computational power, is to process content while it is at an appropriate level in the compression hierarchy. The sequence of potential processing levels is:

- Sequence
- GOP
- Frame
- Slice
- Macroblock
- Block

Although this complicates the process, content quality can be maximized while processing delays can be minimized. Figure 9.22 shows, in a conceptual way, how

Table 9.15 Compression Standards – Features Comparison Chart

	MPEG-2 Video	MPEG-4-Part 10 AVC H.264/AVC	SMPTE VC-1 Based on Windows Media 9
Intra Prediction	– None: MB Encoded – DC Predictors	– 4 × 4 Spatial – 16 × 16 Spatial – I_PCM	– Frequency Domain Coefficient
Picture Coding Type	– Frame – Field – Picture AFF	– Frame – Field – Picture AFF – MB AFF	– Frame – Field – Progressive – Picture AFF
Motion Compensation Block Size	– 16 × 16 –16 × 8, 8 × 16	– 16 × 16 – 16 × 8, 8 × 16 – 8 × 8 – 8 × 4, 4 × 8 – 4 × 4	– 16 × 16 – 16 × 8, 8 × 16 – 8 × 8 – 8 × 4, 4 × 8 – 4 × 4
Motion Vector Precision	– Full Pel – Half Pel	– Full pel – Half Pel – Quarter Pel	– Full Pel – Half Pel – Quarter Pel
P Frame Feature	– Single Reference	– Single Reference – Multiple Reference	– Single Reference
B Frame Feature	– 1 Reference Each Way	– 1 Reference Each Way – Multiple Reference	– Single Reference
In-Loop Filters	– None	– De-Blocking	– De-Blocking – Overlap Transform
Entropy Coding	– VLC	– CAVLC – CABAC	– Adaptive VLC
Transform	– 8 × 8 DCT	– 4 × 4 Integer DCT – 8 × 8 Integer DCT	– 4 × 4 Integer DCT – 8 × 4, 4 × 8 Integer DCT – 8 × 8 Integer DCT
Other	– Quantization Scaling Matrices	– Quantization Scaling Matrices	– Range Reduction – In-Stream Post Processing Control

attributes of a compressed bitstream source can be used to improve the transcoding process.

Methods of processing compressed content are actively sought by the research community. Some compressed domain capabilities, such as logo insertion and MPEG TS splicing, have been successfully implemented in broadcast products.

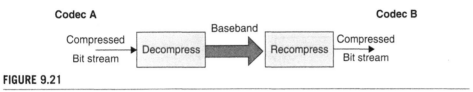

FIGURE 9.21

Brute force transcoding/transrating

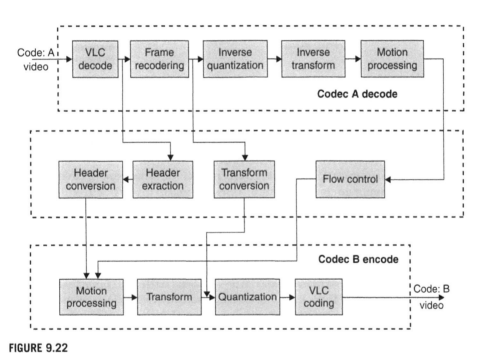

FIGURE 9.22

Compressed domain transcoding

Scaled video coding

Converting and compressing source content for multiple target reception devices can be simplified by adopting a "compress once, use everywhere" philosophy. The technology that can accomplish this is scalable video coding or SVC.

SVC compresses source content in a number of levels of resolution: a base level and augmentation level. The base level is sufficient for complete reconstruction of the original image at a low quality. Each successive level contains higher frequency information—i.e., more detail.

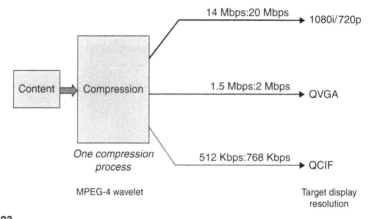

FIGURE 9.23

Scalable video coding

During session initiation, the delivery rate of the channel is negotiated. The level of content to be transferred is governed by this information.

As shown in Figure 9.23, in a three-screen SVC implementation, each delivery channel has a minimum and maximum bit rate. Similar to AVC and VC-1, a single encoder produces multiple bitstreams at different bit rates.

Audio

Audio compression presents a different set of compression challenges. DTV systems can deliver 5.1 channel surround audio in a 384 kbps bitstream. If each audio channel has a data rate of 1 Mbps, a total of 6 Mbps (actually a little less because of the ".1" LFE channel), then a compression ratio of about 15:1 is required to produce the target bitstream. Although it is not the 50:1 required for MPEG-2 video compression, accomplishing the feat and maintaining audiophile quality levels is a difficult engineering task, nevertheless.

Audio compression codecs come in many flavors. For DTV there are basically two: Dolby Lab's AC-3, which is used in the United States for DTV, and MUSICAM, which is an MPEG audio codec that is used in most of the rest of the world. There are also advanced new codecs such as AAC and MP3.

Audio compression methodology

Central to audio compression techniques are the characteristics of audio perception. Psychoacoustics models based on volume, frequency, and temporal masking guide the reduction of unheard audio information, resulting in greatly reduced data rates

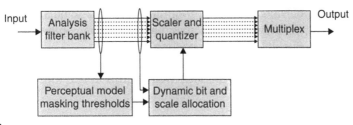

FIGURE 9.24

Generic perceptual audio encoder functional block diagram

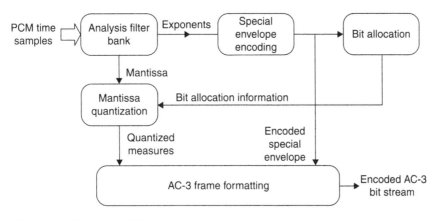

From ATSC A-52 courtesy ATSC

FIGURE 9.25

AC-3 encoder block diagram (From ATSC A/52)

without the lose of audio quality. Volumes have been written about the human audio system; the interested reader is urged to seek appropriate works (Figure 9.24).

Figure 9.24 is generic block diagram of a perceptual audio compression encoder. MPEG as well as other encoders follow a similar processing methodology. Audio is divided into frequency bands based on the characteristics of the ear's critical frequency bands. The resultant data is quantized based on aural perceptual models and the target bit rate.

Coding complexity is structured in three layers: I, II, and III. Mono and stereo at 32, 44.1, and 48 kHz sampling rates results in bit rates ranging from 32 kbps to 384 kbps.

AC-3 and E-AC-3

AC-3, short for audio codec 3, was developed by Dolby Labs for 5.1 cinema sound systems and was selected as the audio codec for U.S. terrestrial DTV. Figure 9.25 shows how the generic audio perceptual codec model has been implemented in AC-3.

This is the encoder block diagram from the ATSC A/52 standard. The perceptual model is not explicitly drawn as a function block. But a perceptual model is used in each of the processing blocks to make coding decisions.

E-AC-3 is an enhanced version of the original AC-3 codec. It includes a wider range of bit rates. AC-3 produces bit rates from 32 kbps to 640 kbps while E-AC-3 has a maximum bit rate of 6.144.

MPEG audio compression

As part of a complete audio and video solution, a number of MPEG audio standards have been developed and implemented on a variety of devices. All MPEG coding is based on psychoacoustic models and uses 32, 44.1, and 48 kHz sampling rates.

The advent of MPEG-1 compression for the first generation of DVDs included audio compression. As part of a complete tool kit, the working group turned its attention to compressing audio for DVDs. The result was defined in ISO/IEC-11172-3, MPEG-1 Part 3.

The next generation, the MPEG-2, audio codec, consists of three layers. Layer 1 is used in DVDs. Layer 2, known as MUSICAM, made a run at inclusion in the U.S. ATSC DTV standard and is widely used in DAB applications.

AAC

AAC is yet another MPEG-based technique for audio coding. Available tools are specified in MPEG-2 Part 7 and MPEG-4 and support multichannel audio. The MPEG-4 tool is referred to as HE (high efficiency) AAC V2; the codec is also called AAC Plus. As the name implies, the goal is to produce acceptable quality audio at very low bit rates.

HE ACC V2 has been adopted by DVB, 3GPP, and 3GPP2 for use on 3G and mobile networks. Audio data rates are 48 kbps or less, well suited for the limited capacity of mobile delivery channels.

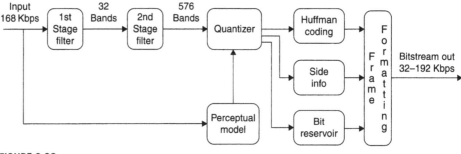

FIGURE 9.26

MP3 encoder functional block diagram

AAC uses a technique known as perpetual noise substitution to remove background noise. Noise is rich in high-frequency information and requires high bit rates for coding. Of course, no one really wants to encode noise. So its removal is actually an aesthetic improvement as well as a way to reduce bit rates.

MP3

MPEG-2 Layer III has taken on a life of its own. MP3, as it is known, has found widespread application in the Internet age. With extremely low bit rates, 32–192 Kbps, and rarely noticeable quality loss, MP3 has become the preferred compression codec for many consumer applications.

Figure 9.26 is a block diagram of an MP3 encoder. Notable features are the double division of source audio into 32 and 576 critical frequency bands and the inclusion of Huffman coding.

CONVERSION SCENARIOS

Now that we've described techniques available in the conversion tool kit, following conversion scenarios from source to each of the three screens will show how they may be applied. Content produced for DTV broadcast will be the source.

Compression ratios and channel capacity

The amount of compression applied to raw content is expressed as a ratio. The most familiar compression ratio is the 50:1 (or greater) figure that roughly describes how MPEG-2 fits 1 Gbps (approximately) 1080i HD into a 20 Mbps MPEG-2 TS.

Because there is no explicit blanking interval in the compressed domain, only the active pixels need be considered. In the following analysis, however, keep in mind that the numbers are approximate and there will be some additional data, such as packet headers, checksums and error correction, and concealment information, that can have a significant impact on payload data capacity.

For example, consider an MPEG-2 TS packet appended with 20 bytes of FEC. Of the 208-byte packet, only 184 are payload. If this packet were transferred over a 10 Mbps channel, it would be 1664 bits of packet containing 1504 bits of data. Conceptually speaking, the payload data rate would be reduced by nearly 10% to 9.04 Mbps.

Internet TV

Applying the 50:1 compression ratio to SD 720 × 480 at 30 Hz (480i) at 166 Mbps results in a bit rate of 3.3 Mbps. It would take a hefty broadband Internet connection and an appropriately sized buffer to enable SD video streaming to a PC. VGA resolution video can be compressed to bit rates that are under 3 Mbps. To get VGA resolution into a 768 Kbps DSL pipe, the video compression ratio must exceed 192:1! By reducing the frame refresh rate to 15 Hz, a 50:1 compression ratio would be adequate for DSL.

Continuing to downsize the pixel grid to QVGA, the data rate is reduced proportionally. For this quarter computer display image, the 50:1 compressed data rate is 747 Kbps. This is barely within the capability of a DSL connection, and leaves little room for header and check sum data.

The QVGA option, however, opens many possibilities as to how to use the available display real estate. Program associated information can be presented with the video, and, more significantly, commercial announcements supporting Web-based transactions could generate revenue for the broadcasters and advertisers. All the necessary technology to support this over the Web is in place, but not by DTV service providers. The implementation of T-commerce over DTV is still in the future.

Cell phone

Acceptable quality video content presentation on cell phones requires sufficient display resolution and decoding capabilities. Even with the amazing advances continually packing more computer power into smaller integrated circuits, the form factor of a handheld device limits the amount of heat dissipation possible and necessary for powerful computation devices to function. But beyond this is the bigger problem of supplying adequate power for a sufficient period of time to enjoy audio and video entertainment on portable devices.

Transcoding video from broadcast compression formats to a format acceptable for a cell phone is challenging. Bit rate and frame type need to be within the capabilities of channel delivery bandwidth and device decoding power.

For mobile video, if the data channel supports 384 Kbps, then application of the 50:1 compression ratio to a QCIF display at 30 Hz will produce a 243 Kbps data stream. For a handheld device, depending on the quality of the display, a 15 Hz refresh rate may be adequate, and the data rate would be halved.

A DTV and PC have the ability to store video frames. This buffer is used to reorder frames from their delivery sequence to their temporal presentation. MPEG and other codecs use techniques that allow encoded frames to reference previous and later frames, only coding the difference.

A cell phone does not have the luxury of having large amounts of memory for implementation of a frame buffer. Therefore, compression techniques that store frames must be those that require minimal frame buffering.

When content is created with a cell phone as the target device, this is not a problem. Contemporary codecs offer profiles that will create frames in a manner that can be transmitted sequentially to the cell phone. Figure 9.27 illustrates a few of the techniques.

There are really only two options. The first is to deliver I frames only. This will consume the most bandwidth but require the least amount of decoder computation,

GOP = 15

60 Hz	I	B	B	P	B	B	P	B	B	P	B	B	P	B	B	I	B	B	P	B	B	P	B	B	P	B	B	P	B	B
30 Hz	I		B		B		P		B		B		P		B		B		P		B		B		P		B		B	
15 Hz	I				B				B				P				B				B				P				B	

When reducing frame rate to 30 Hz or 15 Hz, frame transcoding complexity increases and is problematic due to the fact that frames referred to by P and B frames are removed. Transcoding is computationally demanding.

GOP = 12

60 Hz	I	B	B	B	P	B	B	B	P	P	B	B	I	B	B	B	P	B	B	B	P	P	B	B
30 Hz	I		B		P		B		P		B		I		B		P		B		P		B	
15 Hz	I				P				P				I				P				P			

By reducing the GOP structure to 12 and adjusting the number of I, P and B frames, reducing frame the rate to 30 Hz or 15 Hz, is greatly simplified. Transcoding can be completely eliminated.

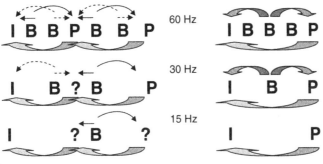

GOP = 15 Reference structure   **GOP = 12 Reference structure**

FIGURE 9.27

Frame rate reduction and GOP issues

conserving device power. The other uses I and P frames; bit rate is reduced but computational complexity is increased.

A factor in determining the best trade off is display pixel grid dimensions and display quality. Can a true HD image be seen on a 3.5 × 2.5 inch screen at a distance of 12 inch or more?

Looking at these figures, it's obvious that a combination of advanced video codecs and increased channel capacity will continue to be developed. AVC and VC-1 have doubled compression ratios for equal video quality, broadband connections tout 4, 6, and 8 Mbps downstream data rates, and telcos are implementing ADSL2+ with capabilities of 24 Mbps and greater.

Converting content to a format and bit rate that is appropriate for the delivery channel and target reception device may be the most important process in the technical preparation of content for multiplatform distribution. But it is just the technical part, the nuts and bolts.

Production, although dependent on technology, is really about artistic and dramatic presentation. While a discussion of the creative process is beyond the scope of this book, possibly even beyond the scope of rational discussion at all, an investigation into the underlying infrastructure that enables producing content for multiplatforms is realistic and the subject of the next chapter.

Multiplatform workflow and infrastructure

10

By now, it is perfectly clear to anyone paying attention in the media industry that digital television is not just about television anymore. With the digitization of content and the advent of nonlinear file-based production, every network and local station has a Web site, frequently more than one. Content is repurposed and distributed to cell phones and handheld devices. And soon mobile television will be available in the United States. Additionally, new outlets such as digital signage in stores, malls, and sports venues—as well as content tailored for elevators, gas stations, and taxis—are increasing the strain on production resources.

In the cost-conscious broadcast engineering industry, the luxury of independent production workflows and infrastructures for each delivery platform—television, Web, and handheld—must eventually evolve to an integrated, adaptable, converged production environment.

With so many distribution channels to produce and repurpose content for, brute-force, linear, independent production workflows just aren't efficient enough. Ideally, content should be properly formatted for each channel automatically—created once, used everywhere. This is what converged production is all about: leveraging one integrated infrastructure workflow to produce content for all distribution scenarios. But this nirvana is much easier to contemplate than to design and implement.

The technical challenges of building an integrated multiplatform converged production infrastructure are unprecedented; transitioning to digital TV transmission has been relatively straightforward by comparison.

In an ideal implementation, workflows for each channel are configured from production building blocks as necessary to meet distribution needs. Elimination of redundant tasks and automation of repetitive processes increases production efficiency, reduces time to air, enables management of multiple versions of content and formats, and maximizes use of infrastructure resources and personnel. To achieve this, workflows are analyzed and documented, and then used in the development of a system conceptual design.

The need to repurpose content for various distribution channels and consumer devices creates a new set of problems not encountered when programming is distributed only over TV distribution channels. Internet and cell phones require different techniques, both technical and aesthetic.

For example, graphics elements cannot be indiscriminately piped through conversion resources, and be expected to result in an acceptable, legible display on all devices. Some fonts and graphics designed for HD will not play well on a 2.5-inch LCD no matter how they are anti-aliased. The Verdana font, for instance, was developed by Microsoft specifically for PC displays; it may not do so well on a DTV or cell phone display.

An intelligently engineered infrastructure married to an efficient multiplatform workflow is the answer. In one MP scenario, appropriate for DTV or NTSC broadcasts, the graphics elements and video are mixed in a PCR with conversion further downstream.

The Chicken or the Egg

The conceptual design process is not a linear workflow. It's hard to know where to begin; but adopting an agile engineering approach, it doesn't really matter. Start with either production workflow or infrastructure conceptualization and document it. Then if you started with the production workflow, review it with the engineering team. If you started with an infrastructure conceptual design, then review it with production personnel. Then do it all over again, in small steps, with frequent reviews.

To use the same content on a cell phone, a different processing workflow must be used. The original video must be converted from a clean feed (no graphics) to the appropriate presentation and compression formats. The graphics must also be converted (key and fill) to a size (and font) appropriate for a cell phone display. Now, there are two parallel production processes.

The Internet offers so many ways to enhance passive TV viewing that it is certainly worth the additional effort to take advantage of all its interactive capabilities and make content consumption a truly "lean-forward" experience. But preparing content for Web distribution will require a third workflow. Graphic elements as well as data elements can be incorporated into an interactive experience.

Content can be produced for all the screens, without having to have completely independent workflows. A converged workflow consisting of parallel production and conversion processes will maximize efficiency: audio, video, and GFX elements are created only once and then converted to the necessary formats for each delivery channel.

CONVERGED CONTENT CREATION

One thing that can be counted on with network programming is that even if the "value" of the content's subject or theme is questionable, the production values are high. Television looks and sounds good, because the people who produce it are good. The production technology is the best available. The artists and on-air talent are exceptional. The producer and director are experienced and innovative.

At the other extreme, content produced for the Web cannot be expected to be of the high production quality found in network TV programs. Content that originates from national cable services and local broadcasters TV will not always be of the same quality as network television; some are equal, a few are better, but most aren't in the same league. Just do a little channel surfing and you'll see what I mean.

Theatrical events, movies, plays, and live musical performances cannot have the same sensual impact as they have in their original presentations environments, even on 70 inch displays with the best 5.1 audio systems. There is no substitute for being there.

User-generated content varies wildly in quality. No one will debate the value of airing a news event captured on a cell phone. The timeliness and "you-are-there" feeling generated by a citizen-journalist captured event has a personalized impact that professional productions do not have, although the pros sometimes strive to mimic it. One issue to consider is how much does it cost to produce broadcast quality content; is it worth the extra expense to attain high production values if no one will ever perceive it? How clean and processed should cell phone audio be for broadcast, or should it always sound like a cell phone?

Time Sensitivity

Animations, snipes, promos, and teases are time sensitive and are inserted into the program stream in master control as the program is disseminated. A rebroadcast on the original platform will not need this information. However, when repurposed, there may be a desire to somehow communicate this information with the intent of driving viewers to the original program offering. For example, the snipe that would run on the bottom of a TV screen runs as a banner on a PC and directs the consumer to the program (or a rerun) on the "big" DTV screen.

A program with a news crawl offers choices. Should the program be archived with the crawl or as clean video? Perhaps there is historical significance to the crawl information, such as on 9/11 when the sequence and timing of information updates in the crawl add to the retelling of the tragedy. Is there a reason to store crawl data as a discrete file? In a sense, it is a news snapshot. Could this information be used by a content classifier system?

Content of lower quality is often presented on a higher-quality device, while higher-quality content is frequently presented on lower-quality devices. Each scenario has its innate challenges. Presenting higher-quality content on lower quality devices is first and foremost a technological issue, second a matter of aesthetics, and third a matter of conveying information. With lesser-quality content, the technical challenge of converting to an aesthetically acceptable presentation on a high-quality device is a tough order to fill.

Repurpose or Prepurpose?

When the opportunity first arose to distribute previously aired content over new media channels, the buzzword du jour was "repurposing"—content was repurposed on the Web, on VoD or whatever.

It didn't take long before a media sage saw the light and pointed out that, since it was a given that content would be distributed over every possible channel, rather than produce content for one channel and repurpose it for others, content creators should "prepurpose" their content while it was in the conceptual stage—that is, plan for its production assembly and distribution over multiple channels.

Workflow strategy

In order to examine the design challenges of converged production for multichannel distribution, consider a hypothetical broadcaster who wants to deliver content over the air, across the Internet, and to handheld and mobile devices. At times, content delivery will be simultaneous; at other times, programming will be available later over nonbroadcast delivery channels.

Obviously, there is a complex interdependence between system resources and workflows. As an experienced broadcast engineer understands, there is more than one way to implement an integrated production infrastructure that supports multiple delivery channels. But no matter how the design challenge is approached, small workflow improvements can have a large impact.

A detailed analysis of how content moves through the production workflow will reveal ingest, storage, and conversion requirements. This should generate a design strategy that works in all scenarios. Try to stick to the fundamental premise that no format conversion is a good conversion, so the number of conversions should be kept to a minimum or zero. Also, storage costs money, and large files in HD formats eat up bandwidth.

Another concern to keep in mind is the SDI routing infrastructure. Plans must be made now for adequate scalability in the future. Can the emerging 3 Gbps format, SMPTE 424/425, be supported with existing equipment? Will implementing faster-than-real-time SDI transfers be required to meet content transfers at peak production times?

Here's a possible initial strawman hypothesis of a workflow: Convert as content is ingested; store in highest quality format, the "house" format; convert as little as possible; and downconvert if necessary only for channel delivery. To maintain the highest video quality whenever possible, however, content may also be stored in its original format.

Essentially, there are two kinds of additional program elements besides audio and video: graphics and data. By dividing program-related material into graphics and data, we can simplify workflow analysis, system design, and development.

When repurposing programs, the task is complicated by the fact that graphics and program-related data have many uses, exist in many different formats, and are presented and/or processed in various combinations. Each reception platform will require a different program assembly technique to present audio, video, graphics, and other data.

Generally, a dedicated and independent infrastructure supports graphics operations. Production of graphics for consumption on devices other than the one originally intended creates workflow and formatting issues that must be resolved.

Presentation problems may arise when porting graphics to reception devices. Sizing issues are of particular importance. For example, will a lower third graphic be visible and legible on a cell phone? When repurposing the content on a PC, would it be best to display the lower third graphic outside of the video window?

MEETING THE MULTIPLATFORM ENGINEERING CHALLENGE

The design of an infrastructure to support integrated production for diverse distribution channels is no simple task. In the broadcast industry, the integration of workflow and technology is more of a mystical art than the direct application of technology.

In a converged production workflow, source audio and video are produced once, in a house production format; then graphics elements are created in a way that is appropriate for each delivery channel. The audio, video, and graphics elements are assembled for each targeted reception device, and then converted to the appropriate presentation and transmission formats and distributed over the desired channel. This integrated approach to production is potentially more cost-effective than storing many copies of the same content in different formats and implementing discrete production workflows to support each delivery channel.

A New Technology Discipline

Broadcast facility engineering departments are now being asked to become experts in the use of equipment designed to support the production and dissemination of content over many channels. Meanwhile, much of the broadcast technical community is still struggling with the integration of IT and computer science into a unified infrastructure. In an effort to adapt, engineering management is trying to help broadcast engineers become more IT-savvy, or they are hiring IT professionals and teaching them broadcast technology and philosophy.

What's really required is a new breed of true media systems engineers—professionals with hands-on and conceptual expertise and years of experience in a wide array of technical disciplines relevant to the media industry. Acting as a team coach as well as project manager, these unique individuals coordinate the work of experienced engineers, IT personnel, application developers, and security specialists by using a combination of traditional linear and contemporary agile engineering project management techniques—concepts we talked about in Chapter 6.

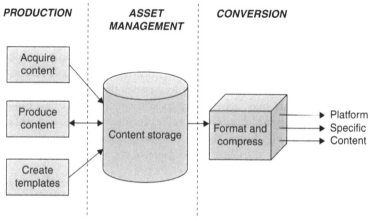

FIGURE 10.1

A converge production workflow for multiple platform content repurposing consists of three fundamental processes: Production, Asset Management, and Conversion

Smarter technology

Converged production isn't about working harder; it's about implementing technology that works smarter. New ways must be devised that enable production control rooms to support multiple distribution channels. Transcoding and graphics assembly can be done on the fly, so when content leaves master control it is properly formatted for both the delivery channel and the target reception device.

The basic multiplatform workflow is deceptively simple. It can be broken down to three core stages as shown in Figure 10.1.

The processes in each stage are straightforward. Acquire the source audio and video. Produce the finished content from the audio, video, graphics, and data elements. Create presentation templates for each channel. Store all the content in a centralized asset management system. When it's time to "take it to air," format the elements for presentation and compress audio and video, packetize the resultant digital data, and hand it off to transmission. As simple as it appears in concept, this is not such an easy task when you consider the number and variety of systems that must work together.

Stepping back to get the total facility overview, Figure 10.2 is a generic conceptual diagram of an infrastructure that supports simultaneous distribution of content over television, Internet, and cellular distribution channels.

In this highly simplified workflow, content is acquired from a studio, a remote production truck, or VTRs or servers via network or satellite content distribution networks. SDI content is compressed as it is ingested to the MAM system. Feed-through sources, such as a live remote event or network program feed, are routed to playout servers. So far it is the same as the usual TV workflow.

A dedicated playout server transcodes and assembles content for distribution over each channel using the appropriate formatting and compression. This is the primary difference between the television and multiplatform workflows in this example facility.

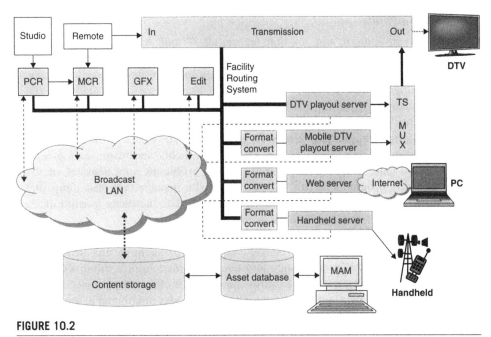

FIGURE 10.2

Functional block diagram of a multiplatform production infrastructure

The three stages discussed above are easy to identify. The studio, truck, graphics, and postproduction are Stage 1. Stage 2 is represented by the two cylinders located in the bottom portion of the illustration: content storage and asset database. Format conversion is Stage 3. At this point in the design process, the infrastructure can be partitioned into functional subsystems.

System solutions

Equipment manufacturers have recognized that multiple distribution channels have to be supported by broadcasters. In response, they are attempting to simplify broadcast system design. Their efforts are producing integrated, configurable devices, with new features and capabilities. Increasing more equipment directly supports the integrated functionality needed to support the three-screen universe.

Since this is a vendor-agnostic book, systems and features will be discussed in a conceptual manner. Actually this is a good thing; it will help to illustrate how engineering is the reduction of a concept to practice. In the future, when designing a system, rather than ask the vendors what they can do and then design to these capabilities, conceptualize what you want a system to do and then find vendors that can provide the building blocks.

That being said, we'll take a moment to discuss two approaches to computer-based equipment design. The discussion applies to both off-the-shelf systems and to equipment that uses an embedded computer.

Hardware platforms

All broadcast equipment, and for that matter, all IT and other electronic devices, depend on the physical technology that they use—the hardware. Printed circuit boards, power supplies, integrated circuits, and discrete components are the physical foundation of electronic systems.

However, since the advent of the microprocessor and other programmable components in the 1980s, the trend has been away from equipment that requires physical modification to implement a new feature, to devices that can be upgraded by modifying firmware. Firmware devices are programmable hardware, such as a field programmable gate array (FPGA). Patches to fix problems, new features, or even completely new programs to run the device can be installed via disk, jump drive, over a network or through a configuration computer. The hardware portion of electronic gear is now a versatile platform.

Compression equipment is one area where this flexibility can be especially helpful. Compression algorithms have evolved rapidly. For example, the best the first generation of MPEG-2 HD video encoders could do for complex video scenes was to produce a bit stream of about 17 Mbps. As research efforts developed improved compression algorithms, a software upgrade to the encoder might enable the encoder to produce a 12-Mbps bit stream with video quality equal to that of the earlier generation of equipment. Today, with the rollout of new compression codecs, software upgrades could enable an encoder to produce MPEG-2, AVC, and VC-1 bit streams as desired.

Since processes are executed in hardware, rather than in software, high throughput speeds are attainable; they are significantly faster than with a software solution.

Software

Software is a world unto itself, and a very complicated world at that. Basically, though, software application implementation comes in two flavors.

The first and oldest is software that uses a computer as a platform. This means that all the functionality provided by the program is designed by the software engineers. The computer is simply a platform. Control of the hardware is explicitly coded by the application engineer; there is no resident operating system, per se. This type of approach enables the design team to have complete control over every aspect of the system they have created.

But we compute in a Windows-based world, whether Apple or Microsoft. Multitasking numerous applications, each in its own window, has become the status quo. In this type of implementation, the application software runs as a process while the Windows OS controls the overall operation of the computer. Software development kits (SDKs) such as Microsoft's Visual Studio make it very easy to design user interfaces: just drag and drop a button and the associated code is created for you; attributes are assigned by filling in fields in context-sensitive menus. This may make life easy, but you are now at the mercy of the SDK and the OS. When something goes wrong during program execution, it may not be the programmer's fault.

PLATFORM-SPECIFIC PROCESSES

Examining a hypothetical multiplatform production workflow will shed light on the design process. First, content presentation formats for each distribution channel need to be established. In this case, the house and transmission DTV format will be 1080i. Content repurposed for the Internet will be produced and delivered as quarter screen 320 × 240 at 60 progressive fps; this pixel grid format enables the user to scale content and maintain acceptable video quality in variable size Windows. Cell phone content will use the QCIF format 176 × 144 at 15 fps. Audio will be full 5.1 surround for DTV, and mono for both the PC and cell phone.

Now consider these production requirements applied for a multiplatform package of sporting events. Source content consists of 10 events that are three hours each. For each of the 10 games, production will create two highlight segments: 15-minute on-demand packages for cable and two-minute versions for the Web. One production workflow will produce the two-minute highlight packages from the raw material on game day. A parallel production workflow will produce the 15-minute versions for on-demand cable. This requires the full game content to be stored for up to two days until the on-demand versions are ready to air. After this, the native resolution games can be archived. The games will also be simulcast live to cell phones (Figure 10.3).

Web streaming will use the low-resolution proxies created by the asset management system during ingest. Production will create additional Web content such as

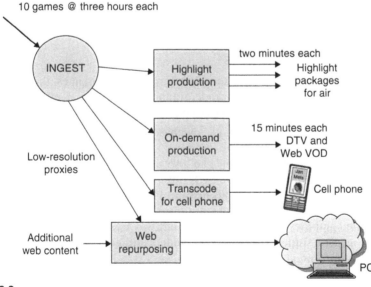

FIGURE 10.3

Hypothetical example of processing content for multiplatforms stats, descriptions, and links. The finished Web versions will be stored for one week on Web servers, after which they can be moved to an archive.

The converged production workflow can be divided into platform-specific requirements. Each of the 10 games broadcast will be produced on site, packaged at the Network Operations Center, and ingested in their entirety.

Cell phone distribution only requires audio and video transcoding. Graphics and commercials are assembled using a technique different from that for the other platforms.

PRESENTATION ISSUES, TECHNIQUES, AND CONSTRAINTS

As discussed in the previous chapter, a format conversion is any change in any characteristic of audio or video presentation. For video, this is any change in combination of display pixel grid dimensions, aspect ratio, scanning method, refresh rate, or color space. In a scenario where content was produced for presentation on a large screen in 720p with 5.1 audio, and then downconverted to QVGA with stereo audio (320 × 240, 2.0) for Internet streaming and CIF with mono for cell phones, the source audio and video formats would be converted twice.

Many choices are available for DTV audio, ranging from 1.0 mono to 5.1 surround. Thought should be given to the production workflow that best produces the correct mix of formats simultaneously for each delivery channel.

Choice of presentation formats must take into account that content will be compressed and then assembled not just for each delivery platform, but also for each decoder. Flash and QuickTime are not interoperable. Neither is MP3 and AC-3. What will happen to a 320 × 240 pixel grid on a 176 × 144 cell phone display?

Template-based GFX processes

Graphics applications often use template-based methodologies that assemble the finished graphics by populating predetermined fields in a display template with data on the fly; sometimes the data is updated in real time. Information that populates the template is contained in a database or data feed (like a sport scores service).

New media distribution channels can also use the data that populates graphics templates for TV presentation. As shown in Figure 10.4, graphics and data can be innovatively integrated into the transmission to create new and interesting features and capabilities when presented on computers and handheld devices.

Automated graphics assembly is an efficient way to assemble graphics just in time to take to air. For sophisticated data-dependent designs, the elimination of manual intervention and integration of database access is a powerful process and speeds up the time-to-air. This is particularly useful in a live production or news show. Production assistants can manually enter data into preproduced templates; they can be previewed and taken to air in seconds.

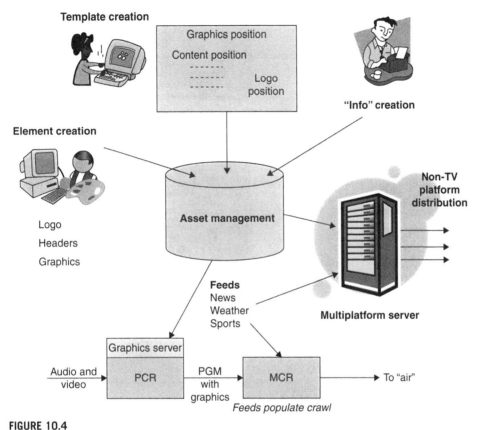

FIGURE 10.4

Graphics template production workflow process

Don't Kill the Goose That Lays the Golden Eggs

If delivered as a discrete data stream, features could be developed that allow the lower third, crawl, clock and score, and other supplemental graphics to be turned on and off by the viewer. The data could be parsed for user input keywords that are used to present only information the consumer is interested in.

An issue with this approach is that TV broadcasters sell sponsorship. Placing these features under viewer control may lessen a sponsor's reach and reduce the price that can be asked for this type of sponsorship.

The workflow is pretty simple. First, graphics templates are created. For example, consider a full-screen, dual headshot with online voting results and previous results displayed in the lower third. The layout with placeholders is created in a graphics application; headshots and previous results exist in a database. Online voting is updated in real time and is periodically updated in the template. An application

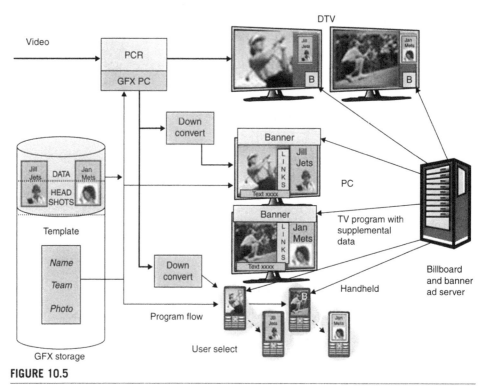

FIGURE 10.5

Multiplatform GFX template population

will direct all this information into a template and push the completed graphics to a graphics playout server, at which point the graphics are ready to air.

Figure 10.5 illustrates the basic template assembly technique used for DTV distribution.

For example, a golf leader board that appears in the center of a TV display can be implemented in Web scenarios to appear in its own Window and include hypertext links to additional information. On a cell phone, the user will be able to drill down for more info.

Graphics conversion

Fonts and graphics must be carefully selected or designed to minimize aliasing when scaling between formats. Should they be produced once, in highest resolution, and then downconverted? Or is it best to create graphics multiple times in the final delivery format, going on the assumption that no conversion is a good conversion or any conversion will degrade image quality?

In any event, good engineering design practice dictates performing an analysis of graphics artifacts produced by conversion as they travel through the production process and distribution chain.

Data formatting and delivery

As more information is displayed in a TV broadcast, and as more of it resides in a database, the process of accessing desired information can be automated for each platform. As discussed in the preceding text, for DTV data populates a template, for a Web page data populates a template or can be accessed by a hyperlink, and for a cell data may populate a template or be an optional drill-down screen.

Sports tickers, stock tickers, news feeds, and weather and traffic information are all examples of data that is used extensively in TV broadcasts and can easily be repurposed for new media platforms.

RSS Feeds

RSS feeds predate the WWW. They are a modern form of the original Internet newsgroups. Since some DTV receiver manufacturers are beginning to implement Web connections, why not exploit them as a delivery mechanism for enhanced features for on-air programming and commercials?

Asset storage and management

Regardless of the workflow and supporting system architecture, the heart of all file-based production is asset storage and management. The fact that content can be acquired in many different formats complicates both the technical process of format conversion and the conceptual workflow design—should content be stored in its original format, or should it be converted to a single "house" format?

This makes storage architecture an important design consideration. Keeping copies of content in multiple formats occupies large amounts of storage, and broadcast-reliable storage can be expensive.

Tracking multiple formats is required if more than one copy is stored. It is important to consider how efficiently the asset management application catalogs the same content in a number of formats.

Proxy generation, used in editing or to aid content searches, may be done by the MAM system or ingest server. Again, this is dependent on the overall system architecture. Proxies, even low-resolution, low-bit-rate images, will need a large amount of permanent storage. Once content is moved offline or archived, proxies and metadata will be the only way to find the source files. Tapes are physical and can be found, but files are bits and are lost without a tag. Proxies must reside in the MAM system forever.

There is a need to examine graphics- and data-related issues with respect to the ingest of audio and video content. Graphics, bugs, closed captions, Program and System Information Protocol (PSIP), and Digital Program Insertion (DPI) triggers require special management techniques. Rather than storing a finished program with composited graphics and embedded closed captions and other data, maintaining these related program elements as discrete but associated items will enable efficient repurposing and redistribution.

As a result, the content management discipline has borrowed concepts and terminology from the computer programming language domain. Collections of audio, video, and other program-related elements are considered media objects. With this content and data in file form, a "wrapper' methodology can be used to associate all elements in the media object.

Compression codecs and multiple source conversion

At this stage in the workflow, the actual transcoding that occurs will depend on detailed production process analysis. The optimal location in the integrated production workflow to transcode must be identified—this will be a key design precept.

The system design goal is to minimize artifacts produced by format conversions—therefore, a fundamental design philosophy is to keep format conversions to a minimum.

Workflow analysis will help decide where format conversions will occur. If a house format is established, conversion takes place at ingest; this conserves storage and eases version management because only one format exists in the production environment. Content is converted as it enters or leaves the MAM, by intelligently located conversion equipment and/or in an integrated production and/or master control room.

The ultimate goal is for the content to be transferred to the transmission equipment in the appropriate format for the target delivery channel; source content may be transcoded concurrently and multiple formats may need to be managed.

Figure 10.6 shows three different approaches to compressing a single content source for distribution to the three device modes. In all three cases, the details of the audio and video process chain are not important, just the sequence of methods.

In Figure 10.6A, a brute force method is employed. One TV source, either HD or SD video or multichannel audio, is distributed to three encoding devices. Each device produces the appropriate bit rate in the desired compression format. The compressed content now moves on to assembly.

A cascaded transcoding approach is illustrated in Figure 10.6B. The baseband source is compressed for DTV transmission. The compressed source is fed to a transcoder and converted to the new bit rate, compression codec, or both. The process can be repeated as often as required. After successive compression cycles, artifacts will be perceived. This is generally a technique to be avoided. However, there may be circumstances where it is the only choice, short of completely decoding a signal to baseband and then compressing it with the new codec. Consider a network TS feed to an affiliate. The DTV feed will be repeated over the local distribution channel. If a mobile broadcast is to be simultaneously delivered, the best option may be to perform a transcode.

The preferred solution, shown in Figure 10.6C, would be to have a single compression encoder simultaneously produce all three bit streams. The practicality of this solution will be difficult to attain in the real world. For example, in the United States, terrestrial broadcasters must use MPEG-2 video compression, but this codec is

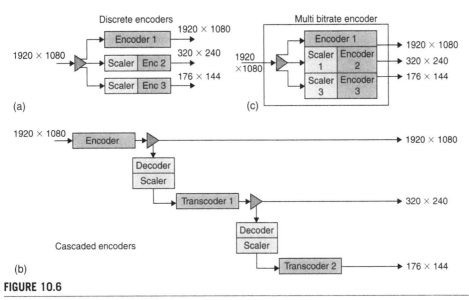

FIGURE 10.6

Three content conversion workflow scenarios

completely inappropriate for Internet or untethered delivery. The bit rate is simply too high.

Many choices have to be made. Dynamic demands on the infrastructure must be taken into account (the best architecture for Saturday afternoon sports production may be completely inappropriate and inefficient for a nightly news program). Another consideration is to take advantage of production downtime. For example, while a two-hour theatrical release is on the air, plan for scheduled maintenance, repairs, and upgrades.

ASSEMBLY STRATEGY

At this point in the workflow flow, the converged workflow diverges. Not only will it diverge based on delivery channel, it will also diverge based on target device. What this means is that many different media players are resident on a reception device. Some can present content in a variety of formats transparently, while others can only present content in a single format.

Figure 10.1, a generic infrastructure that supports multiple distribution channels, left out many details. Figure 10.7 provides the next level of detail.

A server feeds content to the assembly pipeline for all three distribution channels. It can be a single source. It can be multiple sources already formatted for the delivery channel. Or it can be a compressed source, which will require transcoding and/or transrating.

In this scenario, a single stream—say, HD-SDI with embedded audio—is the starting point. The workflow for DTV consists of PCR, MCR, and compression process; these were discussed in Chapter 8.

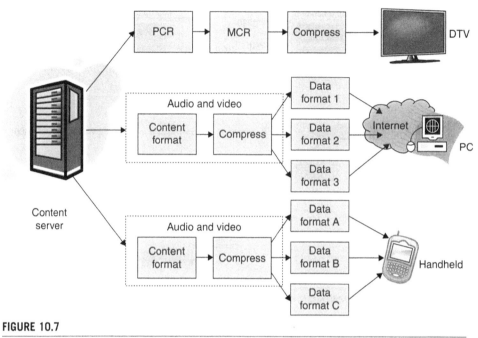

FIGURE 10.7

Assembly systems output content properly formatted for television, Internet, and cell phone distribution

The source content is a program feed consisting of audio, video, and graphics; audio is embedded. In the content format process, the program feed is transcoded to the appropriate format for the intended delivery channel: say QCIF (320 × 240) for the Web and 208 × 176 for the cellular channel. Audio is de-embedded and mixed—say, from 5.1 channel DTV audio to 2.0 stereo for the Web and 1.0 mono for the cell.

Now that content has been properly formatted for the target channels, the compression encoder produces a bit stream that is compatible with the target device. This may be Flash for the Web and SMIL for a DVB-H cell.

The cost of these operations enters into the system design process. Every broadcaster wants to do everything at the highest possible quality, but budget constraints restrict system features to real-world, business model-driven scenarios.

Assembly construction workflow

To assemble audio, video, and graphic elements, planning is important. A simple planning process may include:

- Plan the presentation: determine the features of the delivered content;
- Add elements: create and import media elements, such as images, video, sound, text, and graphics;

- Arrange the elements: on the presentation workspace and in the timeline;

- Apply special effects: such as blurs, glows, bevels, and blends;

- Use scripts to control: how the media elements behave, including response to user interactions;

- Test and publish: be sure the content and features act the way they should on the reception device.

Assembly tools

Content repurposing assembly is done by a graphical interface called a workspace. They may look a little different depending on the vendor, but they all have the same elements in one form or another. These include:

- Menu bar: usually across the top, has pull-down menus to activate commands;
- Tools panel: used for creating and editing images, artwork;
- Control panel: displays context-related options;
- Document Window: primary workspace;
- Panels: aids such as a timeline or color palette.

Figure 10.8 is a generic example of a workspace.

Many popular media players are widely distributed and can be found on all personal computers, cell phones, PDAs, and smart phones. Many are used extensively by video sites, social networking, and business Web sites.

Generally, the assembly tools will produce a playable file that uses a container or wrapper format. A media container, sometimes called a media object, has two parts: the actual media and metadata about the media and encoding used. Most recent versions of the media players support H.264/AVC, VC-1 (Windows Media) video, and MP3 and AAC audio compression codecs. Container formats vary; frequently, they are identified by the file extension. Media files can be very complicated due to the ability to store a container inside a container.

Most assembly applications have the ability to creating content for cell phones and similar mobile devices. The tool kits include scripting language, drawing tools, and templates.

Applications must be thoroughly tested. Some assembly applications can emulate target devices. A developer can select a target device at the beginning of the development process, and then through simulation, be sure to include only those features that the target device can support.

For those who want to control every aspect of the application or just like to do things in the old-fashioned way, tried and programming languages can be used. HTML, JAVA, and SMIL have versions that are specifically designed to support Web and handheld devices.

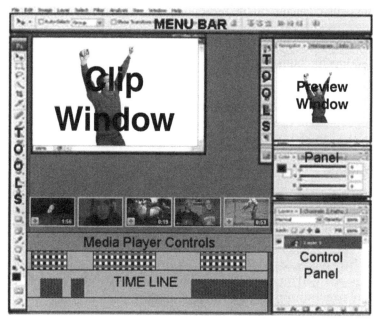

FIGURE 10.8

Typical multimedia content creation application GUI

MULTIPLATFORM PLAYOUT MANAGEMENT

In Chapter 8, we examined broadcast traffic and playout automation systems. This section will explore what is necessary to expand traffic and automation for multiplatform playout. As an example, a television, Internet, and cell phone simulcast will be analyzed. This simplifies the issues; if playout were to happen at different times or be pushed to a server for on-demand access, implementation would become more complex.

First, the system requirements must be established:

- Schedule program playout for each platform;
- Schedule commercial spots for each platform;
- Generate a master playlist;
- Generate automation instructions;
- Move content to channel servers and transcode;
- Manage audio, video, and graphics elements for each channel;
- Create assembly instructions.

Figure 10.9 presents the traffic/automation process.

Fortunately, many of the steps and processes necessary to meet these requirements have been addressed by equipment manufacturers.

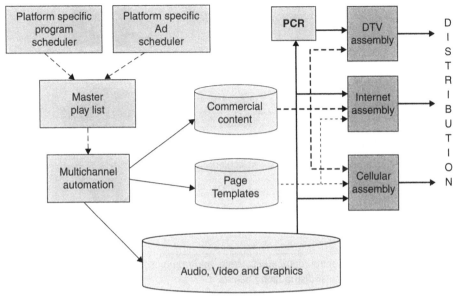

FIGURE 10.9

For a three screen distribution simulcast, master playlist is used by a multi-channel automation playout system to deliver audio, video, graphics, pages, and commercials to the appropriate channel-specific assembly process

Integrated control rooms

Brute force channel- and platform-specific workflows will get the job done, but a more efficient approach is gaining traction. Thanks to the efforts of forward-thinking equipment manufacturers, equipment and systems that can streamline the multiplatform workflow are making their way into broadcast infrastructures.

From a TV-centric point of view, the methods are implemented at two points in the broadcast chain: program control (Figure 10.10) and master control (Figure 10.11).

An integrated program control room

One approach that may lead to increased efficiency and reduce the complexities of distributing content simultaneously over multiple channels is the implementation of an integrated program control room. An integrated PCR, as conceptualized in Figure 10.10, is a control room that simultaneously outputs program elements from common sources in the correct mix for each channel.

One interesting feature the integrated PCR configuration can enable is the insertion of channel-specific commercial announcements. The Web and cell phone release can also include hypertext links, so a viewer can instantly get more information or purchase a product.

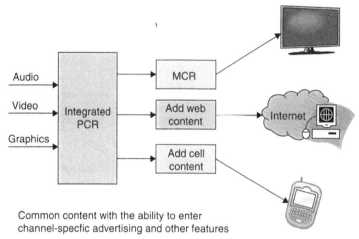

Common content with the ability to enter
channel-specfic advertising and other features

FIGURE 10.10

In an integrated program control room, audio, video, and graphics are mixed and then sent to
MCRs dedicated to a particular distribution channel

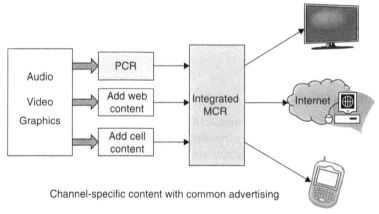

Channel-specific content with common advertising

FIGURE 10.11

An integrated MCR accepts dedicated program feeds from channel-specific processes and
simultaneously inserts commercial content into each channel-specific stream

An integrated master control room

Each distribution channel and reception device has a propensity for content formatted in a particular way that maximizes aesthetic appeal and information communication. Let's consider another control room possibility: an integrated master control
room. Figure 10.11 is a conceptual diagram of a possible scenario.

In this conceptualization, audio, video, and graphics are used as common source
material. For DTV release, the usual signal flow through a PCR and the "integrated"
MCR is maintained. Content for the Web and handheld devices is also passed though
the integrated MCR.

In contrast to the integrated PCR commercial insertion scenario, the integrated MCR can synchronously present commercials across all distribution channels.

Multiplatform Advertising

Along with multiplatform distribution and the concept of "create once, distribute many," the opportunity exists for truly integrated marketing across all platforms. Media organizations can now offer tailored packages over numerous platforms best fit for a particular product. Coupled with targeted ad delivery and t-commerce, the distinction between television and PC functionality is disappearing.

Many media conglomerates that distribute content over multiple channels are breaking from the traditional marketing organizational structure and consolidating their ad sales departments to offer space/time across relevant platforms to align with a client's target audience. The ideal scenario is for publishers and advertisers to have access to the target customer, no matter what media they are consuming.

An integrated program and master control suite

Use of separate control rooms in television has made a lot of sense for a long time. A PCR mixes audio, video, graphics, and other program elements, while an MCR inserts commercials and bugs. In this way, networks could release a program to affiliates, and only an MCR for local insertion of commercials, branding, and bugs was necessary to modify the program for local distribution.

Perhaps the time has come, in light of simultaneous distribution of a program over multiple channels, to view the workflow and need for independent PCR and MCP processes from a different perspective. Figure 10.12 illustrates an integrated PCR and MCR process.

For example, a live broadcast can be produced in the PCR for DTV distribution. If a mix-minus all (or some) graphics is produced, this feed can be used for Web and/or cell-specific applications by adding the graphic content in a way that is best suited for each channel.

Web video looks bad, especially when scaled to full screen. Why not limit the Web Window to less than full screen and present stats, headshots, and crawl elsewhere in a new Window? Let the viewer personalize the crawl, choosing what alerts they want to receive based on favorite team, news topic, or other category.

ENCODING AND ASSEMBLING FOR TRANSMISSION

The demarcation point between the assembly phase and distribution phase of the content life cycle was established earlier for DTV as the production of an MPEG-2 transport stream multiplex. Mobile DTV requires packets to be muxed in the transport stream using a methodology that differs from the way it is done for DTV.

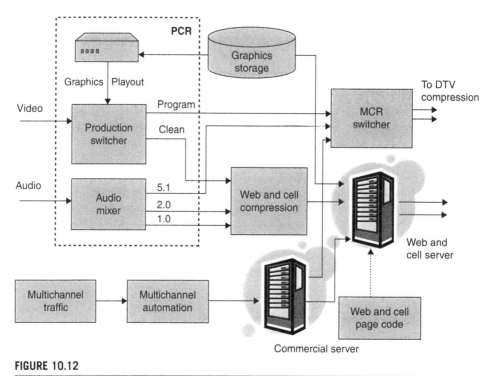

FIGURE 10.12

Functional block diagram of PCR and MCR integration

For an Internet channel, the equivalent point is the format it is in when it is pushed through the LAN gateway to the streaming server; from this point on switching and routing are network functions, ideally transparent.

Figure 10.13 shows the functional signal flow for content after leaving the MCR that will be distributed as an IP packet over the Web or to a handheld device. Compressed content and assembly application (page code) are loaded into the server. In this case, hypertext transfer protocol (HTTP) is used for delivering page content and RTSP for streaming media to the page.

Mobile DTV packetization and transport multiplex

Amazingly, with the variety of DTV system standards used in various parts of the world, the one thing they all have in common is utilization of the MPEG-2 transport stream packet methodology. A similar singularity does not exist for other delivery channels.

One difficult requirement to meet for mobile DTV is that it is truly mobile. With an eye toward deployment in vehicles, systems are capable of receiving and decoding a perfect picture and sound at highway speeds. This robustness comes with a price.

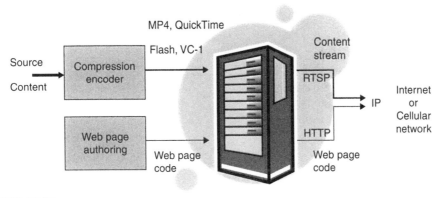

FIGURE 10.13

For Internet and cellular distribution, realtime audio and video and static pages are handled in different ways: pages are downloaded using HTTP and serve as receptacles for streamed media that uses RTSP (Real Time Streaming Protocol)

Forward error correction (FEC) processing is more complex for mobile DTV than for other DTV distribution channels.

In mobile DTV systems under development in the United States, for every bit of DTV payload, three additional bits of error concealment and correction are appended to the data stream. This is a pretty heavy overhead price to pay. A 768 kbps DTV program will occupy 3.072 Mbps of transport stream bandwidth.

Figure 10.14 illustrates the Multi-Protocol Encapsulated Forward Error Correction (MPE-FEC) scheme used by DVB-H, Reed-Solomon RS (255,191) coding (see Chapter 5). This method can correct up to 32 byte errors.

A broadcaster will have to eliminate one of its multicast channels to accommodate a mobile DTV capability. Figure 10.15 illustrates the bit allocation scenario for an MPEG-2 transport stream.

The bits occupied by the mobile DTV signal occupy about a quarter of the bit stream. If an HD program has a bit rate of 12 Mbps and the mobile DTV bit rate is 3.2 Mbps, this leaves approximately 4 Mbps to add a multicast channel and/or data services.

USB Tuners

ATSC USB tuners are the exception to the rule. As a result of increased processing power in very small devices, your PC or laptop can become an MPEG-2 DTV receiver anywhere.

Packet formatting for Internet and cellular transport

Repurposing TV content for cell network delivery is very similar to the Web workflow. The only significant technical difference is compressing audio and video content to a low enough bit rate to fit in the channel bandwidth.

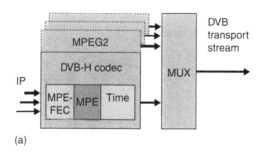

(a)

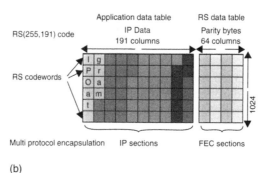

(b)

FIGURE 10.14

DVB Multiplex a) DVB-H codec and transmitter b) MPE-FEC frame structure

In a very real sense, the Web paved the way for cell phones to move from a mobile replacement for landline telephones to multimedia, interactive, handheld personal communications devices.

Multiplatform production workflow automation

The point of process integration is to enable content to be invisibly converted into the proper formats on the fly and with the minimum of human intervention.

Management of multiple versions of content in various formats should be handled by an enterprise application. The correct form should be delivered when requested, by an operator, application, or automation without the need to sift through multiple files.

    NY_v_NE_MPEG2_MP_HL
    NY_v_NE_MPEG2
    NY_v_NE_MPEG2_MP_ML
    NY_v_NE_AVC application determines format
    NY_v_NE_VC1
    NY_v_NE_QuickTime
    NY_v_NE_Flash
    NY_v_NE_Proxy

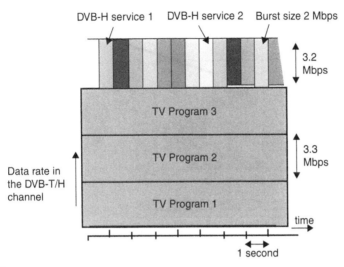

FIGURE 10.15

The time-slicing principle: example of a service multiplex in a common DVB-T/H channel, including time-sliced DVB-H services

Folder-drop automated conversion workflows can be leveraged to assist in content conversion, thereby removing redundant tasks (see Figure 10.16). For example, an editor must produce a segment for DTV, Web, and mobile distribution—a three-screen scenario. The source file is dragged and dropped to a folder designated for such a scenario. The automated compression conversion process kicks in and produces appropriately formatted (i.e., pixel grid, refresh rate, compression codec, bit rate, and color space) files for each distribution channel.

Media "objects"???

Content-wrapping techniques are frequently described as using object models. However, a true object-oriented methodology does more than simply group (or wrap) data together as an object. It also encapsulates operations on the data as part of the object.

The following is a pseudo C++ definition of a Media_Object class that contains a set of member functions that will translate the audio, video, graphics, and data to formats appropriate for DTV, Web, and mobile distribution.

```
Class Media_Object
{
    Audio (channels 1 ..n);
    Video (segments 1 ..n);
    Graphics (lower third, logo, billboard, clock_score, ...);
    Data (PSIP, CC, DPI ...);
    content_DTV Format_DTV (Audio, Video, Graphics, Data)
```

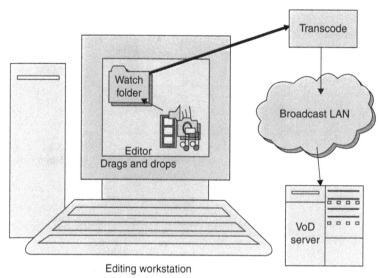

Editing workstation

FIGURE 10.16

Watch folder workflow.

```
    {
        // conversion operations go here
    }
content_MobileDTV Format_MobileDTV (Audio, Video, Graphics, Data)
    {
        // conversion operations go here
    }
    content_Web Format_Web (Audio, Video, Graphics, Data)
    {
        // conversion operations go here
    }
content_Handheld Format_Handheld (Audio, Video, Graphics, Data)
    {
        // conversion operations go here
    }
};
```

LISTING 10.1

Perhaps this is the future direction of content management—the development of content objects that include operations that produce programs in platform-appropriate formats, including graphics and data, with a folder drop or single function call; in other words, a fully automated, integrated production workflow that produces and distributes content over any channel with minimal operator intervention.

Methods will access properly formatted elements. In programming jargon, you'd say that audio, video, graphic, and data elements will be strongly typed based on intended delivery channel.

A HOLISTIC APPROACH

Even though this book takes a TV-centric view of the multiplatform universe, the ultimate goal is the development of a holistic approach to implementation of an integrated infrastructure. Just to be clear: By integrated, I mean an infrastructure that is designed from the outset to create the most efficient and flexible system possible for producing and assembling content for any distribution channel. Sadly, many organizations today just toss content over the wall—to the IT department for Internet repurposing and outsourced to a service provider for handheld content delivery.

Ah, for the simple days when television was just television. Fortunately, broadcasters and broadcast system engineering and design companies are not alone in facing these challenges. The media industry as a whole, from standards bodies to equipment vendors, recognizes where the future is taking us. Hardware systems and software application suites that address integrated production, multiple distribution channels, and multiple reception device requirements are just beginning to appear on the broadcast equipment market.

It is because of the increasing power and adaptability in broadcast equipment that an integrated infrastructure can even be contemplated. So it's helpful to take a moment and discuss the ways in which new technology impacts media systems design and implementation.

11

The interstitial

WHERE WE'VE BEEN

We began by discussing just how the media industry came to this new realm. All three of the delivery channels—television, Internet, and cellular networks—have been around for over a decade. But the combination of increased user uptake and technological innovation has led to this magic moment of alignment and facilitated this new media age. In a very short span of few years, it may be possible that any content at all can be consumed at any time, any place, on any device; however, we're still a long way from the day when ALL content can be consumed regardless of time, location, or device.

This led to a description of just what each of the three device types, the natural extension of the three-screen model, has to offer. What are the user expectations and experience in each domain?

WHERE WE ARE

At this point, a reader should have an improved understanding of how technology, creative and business factors, and goals impact the media industry.

Media systems are deceptively simple on the surface, but increasingly more complex as one gets deeper and deeper into infrastructure details.

The reader should have an appreciation of the need to do a meaningful workflow analysis and a technology process analysis, and then an analysis of how the two fit together. Yes, technologists are analytic when evaluated for communication and personality characteristics at Charm School. Six Sigma and other techniques can aid in developing methods that improve efficiency and increase ROI.

WHERE WE ARE GOING

Technology for technology's sake is fine, but it really serves no purpose or adds any real value to existence. With all the potential that comes with new media, is it fair to ask to what end will it be applied? Will the world be a better place, and will life become easier because of these technological advances? Or will the vast "TV wasteland," described by FCC Chairman Newton N. Minow at the 1961 NAB Conference, expand to all corners of the new media universe?

THE WHOLE SHEBANG

As we've discussed, moving from TV-centric production and distribution to repurposing TV content over new media channels is more than just an exercise in scaling up an infrastructure. There are fundamental differences in all aspects of production and content assembly, both technical and creative, that must be addressed. And of course technology and creativity cannot stand on their own; they must have a basis in a new media business model.

The lesson to be learned here is twofold. First, don't be fooled by simple workflow and functional block diagrams; media technologies are sophisticated. Second, the fact that a media infrastructure is now a mixed bag—a system of systems—pushes complexity to nearly infinity.

This brings us to the second point: As complicated as the foundation technologies may be, as sophisticated as each system is, and as mind-boggling as the entire infrastructure is, they are within the reach of mastery by engineers and technologists. But professional egos must be put aside. It is a stark truth that no one person can understand all aspects of the infrastructure to the technical and practical depth necessary to design and commission a robust and resilient facility. It just cannot be done.

So as the lines of creative, technical, and business distinctions disappear while the industry moves farther into this new universe in the twenty-first century, it is important to remember that we are all in this together. And to liberally paraphrase ol' Ben Franklin, we must all work together in a holistic approach, lest the opportunities before the television and media industry collapse, leaving nothing behind but the confusion and dust—unfulfilled promises of a more connected existence.

Appendix A: Converting between analog and digital

Einstein described space-time as a continuum. There aren't any discrete steps in our conscious perception of the universe. Time flows; we continually age. Sunrise gradually turns night into day. Yes, it's an analog world.

Well, maybe on a subatomic quantum level it isn't, but human perception only goes so far. The resolving powers of the eye and ear have limits. The numeric granularity of digital audio and video exceeds these limits. TV displays and CDs create the illusion of continuous visual scenes and uninterrupted sound because you can't perceive the visual and aural temporal and sensory steps.

A PERFECT DIGITAL WORLD

The transformation of the analog media into a digital representation is the fundamental "enabling technology" of the multiplatform universe. All modern broadcast systems process, store, manage, and present audio and video that has been converted to digital form. Legacy analog formats are digitized and "ingested" into media asset management systems; from that point on, they too are digital.

Many digital systems function like clockwork in a very literal sense. In a synchronous digital system, a periodic signal establishes a moment when data is valid and not changing. As one might logically expect, this signal is called a clock. It may be a sampling clock in analog/digital conversions or a system clock in a microprocessor.

Figure A.1 illustrates the concept, showing the clock signal, represented as a square wave, and data signal. The up arrow in the clock signal is the moment that data is valid. Notice that the data signal is not changing its "level" when the clock signal is transitioning from low to high. This is an important design technique in engineering digital systems.

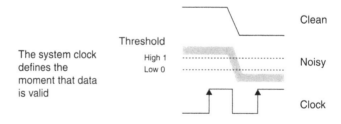

FIGURE A.1

The relationship between system clock and valid data and the effect of noise on digital signals

278

Even a noisy signal (all electrical signals have some amount of inherent noise; it is a natural physical characteristic), as shown in Figure A.1, can perfectly convey a 1 or 0 as long as the amount of noise is not so large as to obscure the logic level thresholds. This is the primary advantage of digital media over analog.

In contrast to synchronous digital systems, an asynchronous digital system does not have an independent clock signal. Valid data is recovered using different techniques that either enable deconstruction of a clock signal or do not require a clock signal at all.

PCM BRIDGES THE DIVIDE

While British expatriate Alec Reeves was working in Paris for International Telephone and Telegraph (IT&T) in 1937, he came up with the idea of digitizing analog voice signals. The result was the invention of analog-to-digital conversion using a technique called pulse-code modulation (PCM). Bell Labs first used IT&T's PCM technique in the secure radio system on which Churchill and Roosevelt talked during WWII. While PCM took decades to commercially exploit, the advent of analog-to-digital conversion was a momentous breakthrough.

Figure A.2 illustrates the PCM technique. The value of a varying analog signal voltage is sampled at regular time intervals. That's the essence of it! But of course, it is not quite as simple as it looks.

Breaking the acronym down, the pulse is the rising edge of the sampling clock. The code is the digital representation of the analog voltage, and modulation is the effect that coding has on the pulse. When represented in the frequency domain it is similar to AM modulation of a carrier wave. This will be covered shortly.

The accuracy of the conversion is dependent on two factors. The first is the number of steps used to code the analog voltage, known as the quantization level.

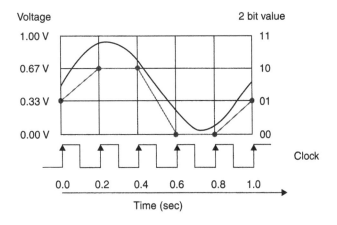

FIGURE A.2

Pulse-code modulation (PCM)

The second is the frequency at which the clock samples data: the more frequent, the more accurate. As seen in Figure A.3, doubling both the number of quantization steps and sampling clock frequency produces a more accurate digitization of the analog signal than the one in Figure A.2.

Digital impairments: aliasing

The frequency characteristics of the source signal determine the lowest possible sampling pulse frequency that will produce an unambiguous, nonaliasing digital conversion of the analog waveform. Harry Nyquist, another Bell Labs research scientist, determined this to be twice the highest frequency component of a signal to be converted.

Figure A.4 illustrates the principle. The waveform to be converted is first low-pass filtered. This removes any high-frequency analog noise (greater than half the sampling frequency) that may corrupt the conversion and introduce aliases as shown in Figure A.4. If a 1 kHz sine wave is sampled at 2.5 kHz, nonoverlapping sidebands are produced on each harmonic of the sample clock, at 5, 7.5, 10 kHz, etc.

But if the sampling frequency is decreased to 500 Hz, the sidebands produced by the modulation of the sample clock will overlap (see Figure A.4b). When this undersampled digitized signal is converted back to analog, more than one signal can be reconstructed from the sample data and aliasing occurs.

If the fundamental sampling clock requirement is not observed, the digital conversion of an analog audio or video signal will be inadequate. This will result in artifacts and less-than-perfect sounds and images (such as a moiré pattern). And of course, it will never do for HDTV and surround sound in a home theater.

Digital-to-analog conversion

Reconstructing the digitized data into its original analog form is more or less the reverse process. Figure A.5 illustrates the steps involved.

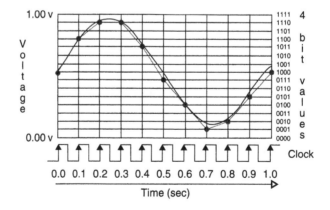

FIGURE A.3

Effect of doubling number of quantization levels and sampling clock frequency on analog-to-digital conversion

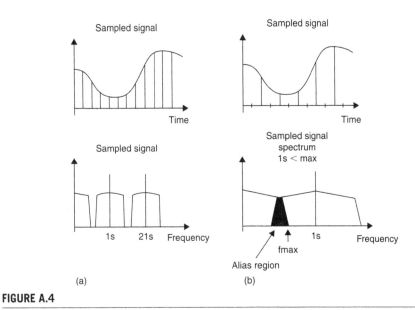

FIGURE A.4

Adequately sampled (a) and undersampled (b) signaling; note aliasing in b

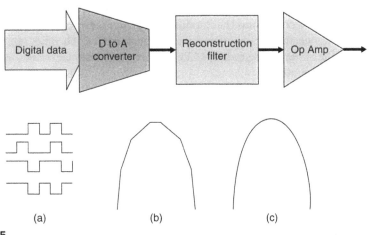

FIGURE A.5

Digital-to-analog processing chain with summing converter and reconstruction filter. (a) Digital input signal. (b) converted waveform, note jagged shape at sampling points. (c) perfectly reconstructed signal after filtering

The digital code words are mapped to an analog voltage level. For example, using 8-bit code words and a 1V range, 0 is 0V, 255 is 1V, and 128 is 0.5V. (Note that the value is slightly higher since there is no true middle value.) This results in a "bumpy" wave because of the low-pass filtering applied before conversion from analog to digital. A reconstruction filter produces a signal identical to the original.

Appendix B: Computer systems

Broadcast operations, content servers, graphics workstations, automation systems, traffic, and billing are computer based. Most nontechnical people have a love/hate relationship with computers. As in all relationships, knowledge of what goes on inside the object of our interest will help foster understanding. This can lead to working out kinks in a relationship; in our case, the entire media business is heavily dependent on what's going on inside computers.

COMPUTING MACHINES

In the late 1930s, researchers were developing the descendents of Charles Babbage's mechanical "Difference Engine." In the process, they realized that electrical technology had evolved to the point where an all-electronic numerical computing device was conceivable and attainable.

John Atanasoff, a physics professor at Iowa State College, was trying to develop ways to facilitate the process of calculating solutions to systems of linear algebraic equations. He became convinced that a digital approach offered considerable advantages over the slower and less accurate analog machines of the time. In December 1939, with the aid of graduate student Clifford Berry, the first electronic digital computer prototype was built.

The prototype included four operating principles: use of the binary number system, regenerative data storage, logic circuits as elements of a program, and electronic (rather than mechanical) data-carrying media.

Let's look at each part.

Boolean algebra and binary numbers

Just as numbers can be represented with digits ranging from 0 to 9—with their location representing the number of 10s, 100s, etc.—so too can other number systems perform the same functions. In everyday life, the 24-hour day and 365-day year are examples of "counting" using something other than 10 as a base. From a defined starting point in time, every 24 hours, the date changes, and when 365 days pass, the year is bumped up by one.

In 1854, George Boole published "An Investigation of the Laws of Thought, on Which Are Founded the Mathematical Theories of Logic and Probabilities." In this seminal work, Boole investigated a binary number system based on two values: 0 and 1, off and on, true and false.

Writing large numbers in base 2 results in very long strings of "1s" and "0s" that are nearly impossible for a human being to understand.

For example, the 16-bit binary number "1010 1100 0010 1001" can also be expressed in hexadecimal numbers. Hexadecimal is a base 16 number system. Each digit from 0 to 9 and A to F represents 0 through 16 while each place is a power of 16.

The conversion of the 16-bit binary number above to "hex" results in: A C 2 9.

The decimal equivalent is $10 \times 4096 + 12 \times 256 + 2 \times 16 + 9 \times 1 = 44073$.

Let's get logical

A natural extension of Boole's work was binary logic. He applied three basic logical functions to binary values and came upon a group of axioms that form the basis of digital logic circuits that Atanasoff utilized in his prototype computer.

Three basic combinations—AND, OR, and NOT—of 0s and 1s in time were supplemented with a fourth, eXclusive OR (XOR). These now comprise the set of basic combinatorial logic functions.

Figure B.1 shows how logical operations are performed using truth tables. The first column is the name of each logic function; the second column shows the symbol used by electrical engineers when designing digital circuits using Boolean algebra. The following columns are truth tables and the logic symbols used on schematic drawings. In the final column, set theory representations of the logic functions are illustrated.

Mathematical functions can be performed by combining these fundamental elements.

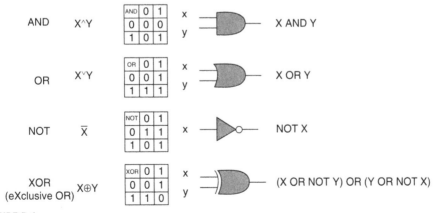

FIGURE B.1

Boolean logic truth tables, electrical gate symbols

Hold that bit!

Regenerative data storage simply means the ability for a circuit to store the value of a bit. The device has an input and output, so the data can be written and read.

Core magnetic memory

In the early days, the storage of programs and data was achieved by winding wires around magnetic cores that were torid in shape. Wires wound in one direction were 0s, while those in the other direction were 1s. This gave birth to the term "Core Memory." The Lunar Excursion Module used in the Apollo space program utilized this technology for storing the program used by its onboard computer.

Catch the bus

The final piece in the original prototype was the use of electronic elements as data-carrying media. At the time of Atanasoff's work, computing machines were using mechanical technology to transfer mid and final calculations. His prototype used wires and electronic signals, rather than mechanical linkages, to convey data.

These electrical interconnections evolved into sophisticated, multilayered circuit boards in early computers, and onto interconnections on system backplanes.

Development proceeds

In the fall of 1937, at the same time that Atanasoff was working in Iowa, Bell Labs engineer Dr. Georges Stibitz used surplus relays, tin can strips, flashlight bulbs, and other items to construct his "Model K" (for kitchen table) breadboard digital calculator, which could add two bits and display the result.

By late 1938, Bell Labs authorized development of a full-scale relay calculator. The design team began the project in April 1939 and, in two years, produced the complex number computer. The device that first ran on January 8, 1940 was connected to three teletype machines and used 400 telephone relays. Lab management felt the technology could aid in the development of long-haul telephone networks.

Theoretical Computing and Turing Machines

While work proceeded on the complex number computer, Alan Turing was developing a rudimentary model that would establish a formal, theoretical computing logic. Turing's work is fundamental in the theory of computation, as all undergraduate computer science majors can painfully attest.

Turing machines, which the mathematician first described in 1936, are abstract symbol-manipulating devices that can simulate the logic of any computer algorithm. Turing machines are not intended as a practical computing technology, but a thought experiment about the limits of mechanical computation.

A Turing machine mathematically models a machine that mechanically operates on a tape on which symbols are written. These symbols can be written or read one at a time using a tape head. Operation is fully determined by a finite set of elementary instructions.

COMMERCIALIZED COMPUTING

WWII accelerated the need for high-speed calculations of artillery trajectory. Engineers at Harvard built the vacuum tube-based Mark I computer in 1944. By this time, the basic architecture of a digital computing machine had been established. Figure B.2 shows the components.

The input and output components enable the machine to communicate with the outside world. Switches and keypunched cards were used for data input, while a teletype machine would print the results of program execution. The memory unit stores data. The processor performs arithmetic and logical operations. The interconnections, or busses, are also considered part of the processor.

Development accelerated, and by 1946, engineers at the University of Pennsylvania demonstrated the ENIAC, the first general-purpose electronic computer. In 1951, John Mauchly and John Eckert build the UNIVAC I, the first commercial electronic computer, which was installed at the U.S. Census Bureau. These machines were still custom-built one-offs.

The first mass-produced computer was the IBM model 650, which the company began manufacturing in 1953. It remained in production until 1969. A total of 1500 units were sold.

All the pieces were in place for computers to make their way into many previously manual business processes. The only problem was that these early computers were huge. They also were expensive. But there was little doubt about their value. As has often been the case, necessity is the mother of invention. It helps to be a little lucky as well.

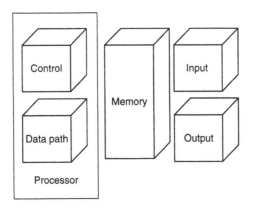

FIGURE B.2

Five components of a computer

SOLID-STATE ELECTRONICS

Electronics was the key to a reliable and accurate computing machine. Mechanical relay technology had been eliminated, but physical size and cost issues still left computer systems beyond the reach of many businesses. Once again, researchers at Bell Labs played a major role in resolving these impediments to technological innovation and proliferation.

Vacuum tubes were fragile and not suited to shock and g-forces experienced during space flight or by missile systems. A new breed of technology was needed for the space race.

Semiconductors

Natural elements can be divided into those that conduct electricity and those that do not. Elements with loosely coupled electrons in their outer valence orbits can easily give up an electron. This flow of electrons makes a substance a conductor. Elements with full outer shells are insulators.

A third group, which includes germanium and silicon, exhibits either characteristic depending on the circumstances. These elements are semiconductors and under certain conditions conduct electricity.

In 1947, John Bardeen and Walter Brattain, working at Bell Labs, were trying to understand the nature of the electrons at the point of physical contact between a metal and a semiconductor. They realized that by making two point contacts very close to one another, they could make a three-terminal device—this became the first "point-contact" transistor.

The breakthrough ignited a huge research effort into solid-state electronics. William Shockley developed a junction transistor, built on thin slices of different types of semiconductor materials that were pressed together. The junction transistor was easier to understand theoretically, and could be manufactured more reliably than earlier prototypes.

Integrated circuits

Size mattered a lot to the Defense Department during the Cold War. Sophisticated electronic guidance systems for guided missiles had to be tough, lightweight, and small. Transistors were an improvement but not the total solution.

In the late 1950s, Jack Kilby at Texas Instruments and Robert Noyce at Fairchild Camera and Instrument Corporation came up with the idea that, instead of making stand-alone transistors, they could make several transistors at the same time, on the same piece of semiconductor. They realized that not only transistors but also other electric components such as resistors, capacitors, and diodes could be made by the same process with the same materials. Dubbed an "integrated circuit" (IC), the method solved the problem of having large numbers of discrete components.

By 1958, Texas Instruments built the first IC. Entire circuit boards could be replaced by a single chip—perfect for the limited U.S. rocket power at the time. This was precisely the solution that the Defense Department was looking for. It is said that during the 1960s, NASA consumed 60% of all ICs produced.

The Real Moore's Law

Gordon Moore, one of the pioneers of ICs and founders of Intel, made an astute prediction that has been applied to just about every type of growth related not only to electronics, but to technology in general. Here's what he said:

The number of transistors that can be fabricated over a given area on a substrate doubles about every year and a half.

Quantum mechanics and the speed of light will cause the doubling to end, theoretically speaking, around 2036, according to a recent IEEE article.

ICs immediately found an additional use in computers. The size of a system was reduced considerably. Chips were designed and produced that could perform logic operations and hold data that previously could only be achieved by circuit boards, greatly reducing the physical size of the machine.

Computer development accelerates

Even with these landmark breakthroughs, computers were still not very operator friendly. Programs were keypunched onto cards that were fed to the machine. Line printers were slow and noisy. System information was presented as a display of lights. Computers needed to become more "personable" and appliance-like to gain widespread commercial acceptance.

Digital Equipment Corporation developed features that were major steps forward in ease of computer use. We take it for granted now, but it wasn't until 1960 that the Programmed Data Processor (or PDP-1) became the first commercial computer equipped with a keyboard and monitor.

The PDP-1's operating system was the first to allow multiple users to share the computer simultaneously. Development continued into the 1980s, with the PDP-11 as the most widely installed model. Amazingly, in the early 1980s, the systems still used wire-wrapped backplanes for printed circuit board (PCB) interconnections.

Computer Time Sharing

Time sharing enables multiple users to connect with a mainframe or centralized computer via a keyboard and monitor. In this way, many users could concurrently use a single expensive and bulky machine.

Ease of use

Even with the user interface improvements fostered by a keyboard and monitor, using a computer was still difficult. Unless you were a professional typist, entering data was slow and laborious. Programs had to be run from a command line. As relatively friendly as computers had become, they still had a long way to go before they would be ready for their coming-out party.

In 1963, the Stanford Research Institute's Douglas Engelbart developed the first prototype mouse. By 1964, this first prototype mouse was interacting with the computer via a graphical user interface (GUI), as well as an early "Windows" system.

Engelbart has said, "It was nicknamed the mouse because the tail came out the end." His version of Windows was not considered patentable (no software patents were issued at that time).

In 1968, the mouse, as well as Windows, hypermedia with object linking and addressing (think HTTP), and video teleconferencing made their public debut. The event was a 90-minute public demonstration of a networked computer system in the Augmentation Research Center at Stanford Research Institute.

Solid-state memory

Today an electronic circuit can be constructed using logic gates that perform memory functions. Figure B.3 shows the internal construction of a D flip-flop from basic logic gates.

The circuit symbol is on the upper left, with the D-latch truth table below. A schematic diagram of the arrangement of logic gates is on the upper right. The lower right is the timing diagram and truth table. The data signal enters the device on the "D" input, while the clock signal enters on the "CLOCK" input. "Q" represents the output data. As is shown, data is valid and "latched" on a rising system clock edge and the "Q" output assumes this value.

Operating systems

Communication between computer components and allocation of hardware and software resources are managed by an operating system (OS). An OS can be conceptualized as concentric circles (see Figure B.4). The BIOS (basic/built-in input output system) or "kernel" is the set of core programs that handles all communication with physical devices—keyboard, monitor, printer, etc. The outermost circle is the user interface. A properly designed OS only allows communication between adjacent layers.

Today, just about every application runs in a windowed environment: multitasking application threads, each in a dedicated Window. But the original, command line interface lives on in professional applications. Configuration of equipment, such as broadcast signal routers and network switches, is still often performed through a command line Window.

Command line programs can bypass the Windows OS and enable direct access to the machine's capabilities. This results in faster, more reliable program execution.

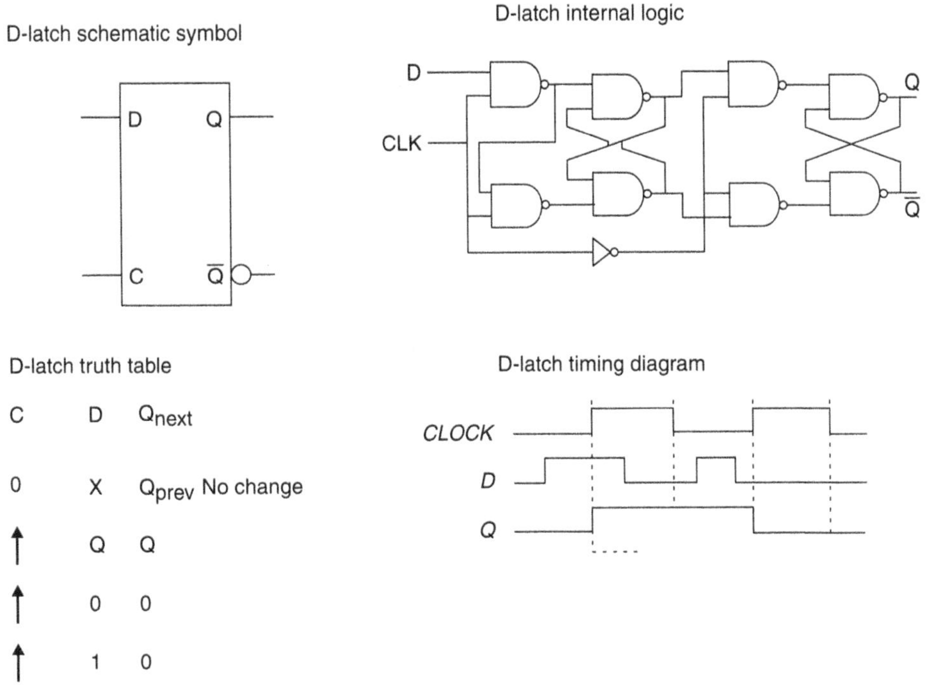

D-latch schematic symbol

D-latch internal logic

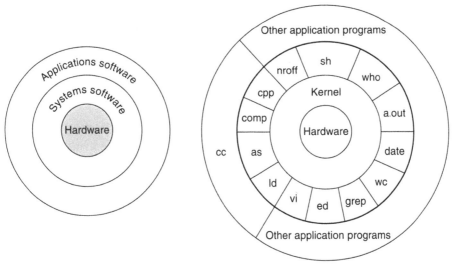

D-latch truth table

C	D	Q_next	
0	X	Q_{prev}	No change
↑	Q	Q	
↑	0	0	
↑	1	0	

D-latch timing diagram

FIGURE B.3

A "D-latch" can be used as random access memory

FIGURE B.4

Layered architecture of computer systems

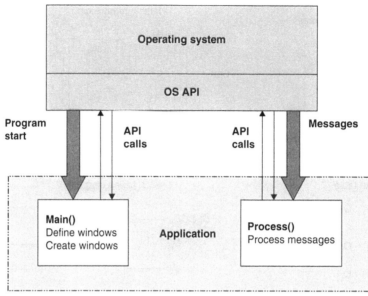

FIGURE B.5

Command line and Windows OS communication

A windowed OS creates a layer between the program and the OS and manages the interplay between the program and the OS. Figure B.5 contrasts the two methodologies.

Modern OS run many "services" in background. This can be why you sometimes experience sluggish application performance. Removing or disabling services that are not necessary can facilitate improved performance. In a broadcast environment, 100% reliability is required.

COMMODITIZED COMPUTING

One bit of Americana associated with the 1980s is the advent of personal computing. What had been a small, techno-hobbyist pastime jumped into the living room. Again, this was enabled by advances in IC design and manufacturer and by the fact that IC prices were falling as economies of scale were trickling down to the consumer domain.

Up until the PC revolution of the early 1980s, computers were primarily large electronic systems dedicated to a specific application; a system was tweaked for its intended purpose. In an era of limited capabilities, as compared to today, every byte of storage and every machine clock cycle was a precious commodity.

In 1971, Intel produced the first microprocessor, the 4004. A general-purpose processing unit designed for desktop calculators, the original device did 4,000 operations a second and used 4-bit "nibbles" for data transactions; hence, the "4s" in its name. The next generation moved on to 8-bit bytes and doubled its performance metrics. Can you guess what it was called? That's right, the 8008.

The 8008, constructed from 3,500 transistors, was slower in terms of instructions per second (45,000-100,000) than the 4-bit 4004 and 4040. But because the 8008 processed data 8 bits at a time and had more RAM, it produced a significant speed improvement in executing many applications.

Personal computers made the scene in 1975 with the first widely marketed product, the Altair 8800. Early personal computer "kits" were the Altair kit and Timex Sinclair kits. Commodore Computers introduced the first computer designed for home users. The VIC 20 (VIC stands for video interface chip) cost $300 in 1980.

Xerox PARC

Although lost in the modern era of Microsoft and Google, the Xerox Palo Alto Research Center (PARC) introduced many of the features taken for granted at all levels of computing from supercomputer to PC. The mission—create "the architecture of information"—began in July 1970.

The Alto, which debuted in 1973, introduced the mouse, computer-generated color graphics, a GUI featuring Windows and icons, the WYSIWYG text editor, Ethernet, and Smalltalk, a fully formed object-oriented programming language.

(WYSIWYG—"what you see is what you get"—described the ability for a computer to actually print exactly what you saw on the display while using a word processing application.)

Other firsts:

1972—Smalltalk, the first object-oriented programming language is invented with an integrated user interface, overlapping Windows, integrated documents, and cut and paste editor; the precursor to C++ and Java programming systems.

1973—The Superpaint frame buffer records and stores its first video image. A decade later, Xerox and its inventor wins an Emmy award for the technology.

1973—A patent memo for a networking system uses the term "Ethernet" for the first time. A few months later, an entry about Ethernet in a researcher's lab notebook reads: "It works!"

1973—Client/server architecture is invented. This development makes the paradigm shift of moving the computer industry away from the hierarchical world of centralized mainframes—that download to dumb terminals—toward more distributed access to information resources.

The sleeping giant awakes

With others having primed the pump and shown how there might be something to this PC stuff after all, IBM management speculated on the potential profitability of moving into the home computer business. IBM's core business was mainframe business computers. There was a lot of debate within the organization as to whether the PC market was worth risking time and money on. This created a rift through management. One could argue that this rift was at least in part responsible for the DOS nonexclusivity blunder.

In 1981, IBM introduced the IBM PC featuring the MS-DOS operating system: MS for Microsoft, a fledging company at the time. The first machines used the 8-bit Intel 8088 microprocessor. Onboard read only memory was a whopping 640 kilobytes, and system clock speed was a lightning fast 8 MHz. Despite all the corporate turmoil, the IBM PC was a success.

Under the hood

A modern PC is a technological marvel, regardless of the fact that everyone takes it for granted. All the hardware used in the machine can be described in a simple illustration (see Figure B.6).

The microprocessor is in the upper left corner. It communicates with advanced graphics processor (AGP) over the PCI North Bridge. The North Bridge connects to the PCI/ISA South Bridge, which in turn communicates the dynamic memory (controller DMA). PCI devices are plugged into the bus between the North and South Bridges, and EISA devices connect to the South Bridge. Simple, isn't it?

Hardware is the platform that the OS and applications run on. If the hardware is the soul of the machine, the heart is surely the OS.

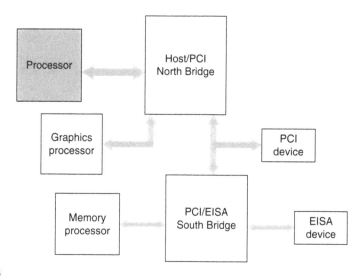

FIGURE B.6

The architecture of a Pentium PC (add peripheral devices)

DSP

Microprocessors were intended to be general-purpose devices. The same chip could be used in a calculator or a control system. All that needs to be done is to write a new software application.

But audio, video, and graphics processing requires certain algorithms to be executed repeatedly. The difference between doing this in hardware rather than in software increases execution speed by 50 times or more. Real-time systems require this level of processing speed.

Digital signal processors (DSP) and graphics processing units (GPU) are microprocessor-like devices that include certain relevant, often-repeated processes as built-in hardware circuits.

Multicore processors

Personal computers and their professional counterparts, workstations, have recently entered a new era. Scientific computing has the need for tremendous amounts of computational power. Supercomputers met this requirement by integrating the operations of multiple processors. The SGI origin series inherited the Cray interconnect and could coordinate the operation of over 64,000 processing cores.

Of course, these were very expensive machines; a 16-processor Origin 2000 could cost just shy of a half million dollars in 1999. So, attempts were made to coordinate the operation of less expensive machines. A Beowulf cluster of Sun workstations is a representative example. It still wasn't cheap, and performance and reliability were problems.

Necessity being the mother of technological innovation, it wasn't long before microprocessor design began to incorporate techniques used in clustering to produce a discrete device with multiple processing cores. Figure B.7 shows a single core microprocessor and three methods of implementing a multicore chip.

Today, multicore processing computers are everywhere. High-end graphics workstations will use 8 cores or more and dual duo, 4 cores. PCs are less than $1000.

By the numbers: number crunching

Computer manufacturers present a plethora of specifications and performance benchmarks all intended to dazzle the potential client. How do you determine what numbers matter? The truth is that individual subsystem metrics only partially define a system's performance. Computer systems are a synergistic relationship of constituent hardware and software components. Poor hardware design, inefficient software, and the failure to properly integrate the two will minimize any theoretically attainable performance capabilities based on subsystem performance specs.

All computers consist of the same basic components: a processor, storage, cache, network interface card (NIC), video/graphics cards, audio cards, keyboard, mouse, monitor, power supply, and everything else connected via USB, 1394/FireWire, RJ-45 network interface, or PCI cards.

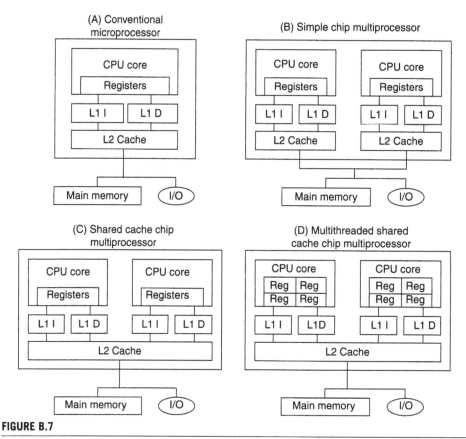

FIGURE B.7

Architecture details of single and multicore processors. Note separate L1 cache for instructions and data

Processor speed is important for number crunching. However, only when all data resides in the CPU is instruction execution performance directly related to "clock" speed.

Mega floating point operations (MFLOP) ratings try to give an indication of platform computational performance. The numbers have some relevance if all calculations are done with floating point numbers.

Front side bus speed will determine how fast data gets to the CPU from RAM: the more the RAM, the more the data blocks that can be read from the disk. Therefore, disk speed influences data transfer rates and consequently overall performance.

Chipset compatibility and suitability must be considered. And hard drive disk speeds and data access times may have more of an impact on the performance of some applications than any other system component.

Even benchmarks provided by testing services should be taken with a grain of salt. It is good engineering practice to test the limits of a machine before purchase.

This "stress" testing can ensure that the contemplated computer will really do the job. Nothing tests a system like real-world conditions.

OS Patches, Service Packs, and Upgrades

Everyone who owns a computer is aware of the need to periodically install patches, service packs, and, occasionally, a full OS upgrade. Although all of these tasks are done with the best intentions of fixing and improving systems, they can also have unintended side effects. In mission-critical scenarios, this risk is unacceptable.

A methodology that has been developed by some high-end system providers is a choice between placing software installation in feature or maintenance mode. Feature allows a full package to be installed that will fix bugs and also install new features. Maintenance mode will install only patches. The choice is made any time a new OS is installed or fully upgraded. This minimizes the risk of side effects popping up at the least opportune moment. Remember, Murphy was an optimist!

Appendix C: Programming languages and application development

With computer-based systems everywhere in the media infrastructure, it goes without saying that, since computers can't tell themselves what to do, someone has to instruct the machine as to how to behave. Media resides on ingest and playout servers; rundowns are written in newsroom applications that communicate directly with wire services, the Internet, and other resources; graphics production and playout are totally dependent on computer applications; and content can be automatically moved to playout servers by automation systems.

THE TOWER OF BINARY BABBLE

Writing code for computer systems is an esoteric art. It has evolved from handwritten instructions that had to be meticulously converted from mnemonics to strings of ones and zeros, and then loaded into program memory by setting switches that were set to the binary code, word by word, and then the pressing of a button. Then the next instruction was loaded in the same manner, and on and on...

Today, designing and coding robust Windows-based applications is a challenge, even to the most experienced programmers. Software development kits (SDK) and module libraries have eliminated the need to write many functions, but with the abundance of source code to choose from, an application developer must make selections carefully. Choose the wrong previously written code and you may spend more time debugging it than writing it.

Language evolution (and generations)

Programming language evolution is described in generations, the first generation being binary ones and zeros. The second generation is known as assembly language, something only an electrical engineer or mathematician could understand and love. Third-generation languages are human readable—just about any logical person can write programs with it (often poorly). BASIC, Pascal, and C are considered third generation.

The first leaps

Grace Murray Hopper, a 1934 mathematics PhD scholar from Yale, joined the Navy Reserve at the beginning of WWII. Her first assignment was under Commander Howard Aiken at the Bureau of Ordinance Computation at Harvard University. There she became the third programmer of the Mark I computer.

At the end of the war, she remained at Harvard as a civilian research fellow in Engineering Sciences and Applied Physics. She left to join Eckert-Mauchly Computer Corporation as a senior mathematician.

Hopper remained with the company when Remington Rand bought it in 1950, and later when it merged with Sperry Corporation. During this time, she developed the first assembler, A-0, which translated symbolic mathematical code into machine code. Using call numbers, the computer could retrieve subroutines stored on tape and then perform them.

The A-2 became the first extensively used compiler, laying the foundations for programming languages. Hopper succeeded in developing the B-0 compiler, later known as FLOW-MATIC, which could be used for typical business tasks such as payroll calculation and automated billing. Using FLOW-MATIC, she taught UNIVAC I and II to understand 20 English-like statements by the end of 1956.

Formula translation

Led by John Backus, a team of programmers at IBM developed FORTRAN (FORmula TRANslation), first published in 1957. It was designed to simplify translation of math formulas into code.

Used extensively in scientific applications, FORTRAN was a third-generation language, but was the first *high-level* language, using the first compiler ever developed. Prior to the development of FORTRAN, computer programmers were required to program in machine/assembly code, such as Hopper's A-0.

The design objective was to create a programming language that would be simple to learn, suitable for a wide variety of applications, *machine independent*, and allow complex mathematical expressions to be stated similarly to regular algebraic notation. It was easier to code FORTRAN than to use assembly language, and programmers were able to write programs 500% faster than before, while execution efficiency was only reduced by 20%.

From Human Readable to Machine Executable

An interpreter translates the program code at runtime into machine instructions.

A compiler translates the high-level program code into assembly language.

An assembler translates assembly language into machine language, a series of 1s and 0s, and loads them into program memory (Figure C.1).

1956: News Flash—Computers Play Chess

The MANIAC I becomes the first computer program to defeat a human opponent in a game of chess.

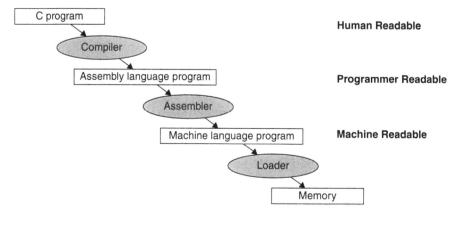

FIGURE C.1

Steps in program design through storing in system memory

In 1959, the first specifications for the programming language COBOL (COmmon Business-Oriented Language) were developed. Another third-generation programming language, COBOL, was initially created by The Short Range Committee, organized by Charles Phillips of the United States Department of Defense. The specifications were inspired by the FLOW-MATIC language and the IBM COMTRAN language invented by Bob Bemer.

Back to haunt them

Bell Labs continued to investigate digital technology and computation-related subjects, regardless of the fact that they were limited in the extent to which they could capitalize on the commercial implementation of any of their work.

In 1972, Dennis Ritchie at the Bell Telephone Laboratories developed a general-purpose computer programming language called "C" for use with the UNIX operating system. Although C was designed for implementing system software, it rapidly found use as an application development language.

The original version of C was a loosely typed language with features taken from third-generation, high-level languages such as PASCAL and easy access to hardware enabled by assembly language.

C syntax has been used to specify digital television systems standards. It has been extensively used by MPEG, ATSC, and DVB (Tables C.1 and Table C.2).

Table C.1 Use of C Pseudo Code for DTV System Specification

Syntax	No. of bits	Format
next_start_code[] {		
while (bytealigned[])		
zero_bit	1	'0'
while(nextbits[]! = '0000 0000 0000 0000 0000 0001')		
zero_byte	8	'00000000'
}		

Table C.2 ATSC Private Information Descriptor

Syntax	No. of bits	Format
ATSC_private_information_description[]{		
descriptor_tag	8	OxAD
descriptor_length	8	uimsbf
format_identifier	32	uimsbf
for ($i = 0$; $i < N$;$i++$) {		
private_data_byte	8	bslbf
}		
}		

Spaghetti

```
100 print Hello
101 subroutine 200
102...
...
200 x = 2+3
201 return 102
```

Structured

```
int x
function add(int a,b)
            return(a+b);
main()
{
        print "Hello";
        add(2,3);
    ...
    }
```

Object Oriented

```
Class Math                    main()
{                             {
public:                           cout<<"Hello";
    int x;                        Math::Add(2,3;)
    int Add(int a,b)          ...
};                            }
int Math::Add(int a,b)
{
    return(a+b)
};
```

LISTING C.1

Spaghetti code, structured programming, and object-oriented programming

Object-oriented programming

There were obvious problems with structured programming. Data and the functions that modified were independent and had to be explicitly linked and called by the program. This and other poor programming practices led to "spaghetti-code" where logical program flow was ignored. Consequently, applications ran slower than they should and crashed (Listing C.1).

C++

Researchers at Bell Labs weren't finished yet. In 1979, Bjarne Stroustrup began work on C with Classes. His goal was to enhance the C programming language. Originally he had been assigned the task of analyzing the UNIX kernel. As luck would have it, UNIX was written in C.

When working on his PhD thesis, Stroustrup found that other contemporary languages had features that were appropriate for large software development, but were slow; others were fast but too low-level to be suitable to large-scale software development. This led him to the idea of enhancing the C language with Simula-like features. C was chosen because it is general-purpose, fast, portable, and widely used. (*Probably didn't hurt that it too was developed at Bell Labs.*)

In 1983, it was renamed C++. Enhancements started with the addition of classes, followed by—among other features—virtual functions, operator overloading, multiple inheritance, templates, and exception handling. The first commercial release occurred in October 1985.

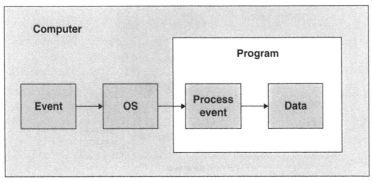

FIGURE C.2

Event-driven processing

Event-driven applications

A technique first implemented in the design of the UNIX operating system has been expanded to modern, multitasking, Windows-based systems. The idea is that small programs, called "daemons" or "listeners," run in the background waiting for "events" to occur. When a relevant event happens, the daemon/listener initiates an action and the "handler" takes appropriate action. Figure C.2 illustrates the concept.

Using this technique, a programmer never really has to check to see if a key has been pressed, or an icon clicked. It is all taken care of by hidden processes.

Java

Java is an object-oriented programming language designed to be OS independent; that is, it is portable. Applications written in Java can be run on any computer platform with a Java Virtual Machine (JVM).

A JVM interprets compiled Java bytecode for a computer's hardware platform. The JVM serves as a translator between the general-purpose Java code and the machine-specific instructions required to actually execute the operation.

The JVM specification defines a generic "machine" rather than the one that is platform specific. The specification includes an instruction set, a set of registers, a stack, a "garbage heap," and a method area. (Methods are analogous to procedures or functions in other languages.) Figure C.3 is a functional illustration of a JVM.

Java programs are written using an SDK and are compiled to an intermediate form called bytecode. Bytecode is a series of bytes that are an intermediate, machine-independent representation of the original Java code. The JVM translates the bytecode to machine instructions in either of the two ways. The first is

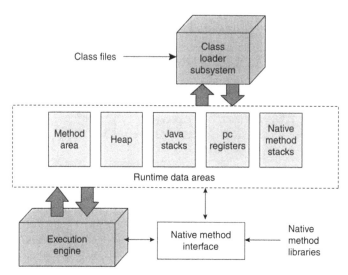

FIGURE C.3

The Java virtual machine

interpreting each bytecode as the program executes. The other is to compile the entire program "just-in-time," that is, before execution begins.

Many media-related applications are written in Java because of their portability. In particular, interactive TV programmers have developed a Java-based SDK.

Web Services

It was only a matter of time until Web browsers and the techniques used for communication on the Internet were adapted to other programming scenarios. After all, every computer has some form of Web browser and the means to interact with other devices over a network.

The result, Web Services, leverages the HyperText Transfer Protocol (HTTP) and a variation of HTML called XML (eXtensible Markup Language). A brief discussion of Web Services is necessary as a foundation for some topics in later chapters of this book.

Web Services are built using three components:

- SOAP (Simple Object Access Protocol)

- UDDI (Universal Description, Discovery and Integration)

- WSDL (Web Services Description Language)

The UDDI is a registry of Web Services described using WSDL and provides a method for publishing and finding service descriptions on the Web. UDDI is XML-based.

All messages are sent using SOAP. SOAP provides the envelope for sending the Web Services messages and generally uses HTTP.

WSDL uses XML to define messages. XML has a tagged message format. Listing C.2 compares HTML and XML.

```
<H1>Paintings by Philip J Cianci</H1>     <gallery>
<TABLE>                                    <artist>Philip J Cianci</artist>
 <TBODY>                                   <paintings>
 <TR>                                       <title year="1995">The GA @ the ATTC</title>
  <TD>The GA @ the ATTC</TD>                <title year="1995">Unloading the Truck</title>
  <TD>Unloading the Truck</TD>             <title year="1994">PCB #121</title>
  <TD>PCB #121</TD>                        <title year="1997">Fremont Street </title>
  <TD>Fremont Street</TD>                  </paintings>
 </TR>                                     </gallery>
 </TBODY>
</TABLE>


HTML: Addresses Presentation              XML: Addresses Data Conveyance
```

LISTING C.2

Comparison of HTML and XML

Figure C.4 illustrates the Web Services process. A service provider published its service to a directory of services using WSDL. The directory uses UDDI. A service consumer checks the directory to locate a service and the communication protocol. The service consumer sends a request to the service provider. The service provider responds to the service consumer.

The big picture looks pretty simple, but a lot of communication, query/response, and handshaking are going on. The process is not as efficient as using a dedicated application. But the fact that all the computer platforms include Web browsers enables interaction between different operating systems.

In the diversified computer environment found in a broadcast infrastructure, interoperability is the highest priority. Many vendors use Web Services for access into their devices for configuration, control, and command operations.

Relational databases

Relational databases is another modern data management technique that is of great importance to the media industry, just as it is to the business world. Relational databases are heavily dependent on media management. We'll discuss them now, because they will begin to appear in functional diagrams in a short while.

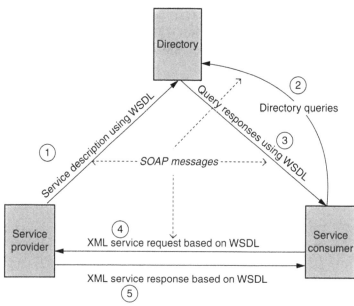

FIGURE C.4

Web Services architecture

American Airlines installed the first large database network, SABRE, in 1955. It was built by IBM, and connected 1,200 teletypewriters. In the 1960s, two main data models were in use: network (CODASYL) and hierarchical (IMS).

E.F. Codd, an IBMer, proposed a relational model for databases in his seminal paper "A Relational Model of Data for Large Shared Data Banks" published by the ACM in 1970. His breakthrough was disconnecting the logical organization of a database from the physical storage methods.

In essence, this abstracted the details of the physical implementation of the data in system memory. This allowed programmers to deal with the contents of the database logically as they wrote applications. They no longer had to count bytes to find the location of an item (Listing C.3).

Listing C.3 contains excerpts of a typical database implementation written in MySQL, an open source database programming language. The three lines that begin with "mysql" are a .bat file that executes the build module, and then the modules that populate the tables. The first lines of code after the "drop" command create the tables in the database.

```
mysql-h localhost <si_hd_archive_1.sql
mysql-h localhost <si_hd_mailings_1.sql
mysql-h localhost <si_hd_mail_info_1.sql
drop database if exists si_test_1;
create database si_test_1;
use si_test_1;
create table mailings
(
     mailing_ID char(4) not null primary key,
     date date not null,
     text_file char(128) not null,
     photos char(128),
     desc_1 char(128)
     desc_2 char(128)
);
create table contents

(
     item_ID int unsigned not null primary key,
     title char(128) not null,
     author char(128)
     proj_cd char(6),
     org_cd char(32),
     type char(6) not null,
     subject char(128),
     info char(128)
     mailing_ID char(4) not null
);

Item_ID, title, author, mailing_ID,
1,"Grand Alliance Board Descriptions","ATV Dept","M_0",
2,"Lecture outline presented at Philips Labs", "ATV Dept","M_1",
3,"NBC Specs pre ATRC & GA","NBC","M_1",
4,"SMPTE Article","D. Fink", "M_1",
...
mailing_ID,date,Text_file,photos,desc_1,desc_2
"M_0","2002/8/22","SI_mail_0.pdf",NULL,NULL,NULL
"M_1","2002/9/21","SI_mail_1.pdf",NULL,NULL,NULL
"M_2","2002/11/23","SI_mail_2.pdf",NULL,NULL,NULL
...
```

LISTING C.3

Excerpt of typical database implementation written in MySQL

Appendix D: Computer networking

A computer on its own is only as powerful as its hardware and software; connect more than one and the power increases exponentially, often characterized as 2^n: where n is the number of machines in the network. The number increases rapidly—2, 4, 8, 16,.... to 1,024 for 10 computers all the way up to over 10 million for 30. Now think of how many machines are connected to the Internet; the power figure is beyond comprehension.

The first computers were stand-alone devices, but it wasn't long before many people envisioned connecting computers together into a network as the next logical development in computer science. Interconnecting computers to enable communication and data transfers among devices was the subject of high-level research.

Today, we take networking for granted. Any consumer electronics store offers a wide array of network switching and routing equipment. One can choose between wired and wireless connections. With autodiscovery capabilities and networked set-up applications resident in modern OS, network configuration is often a plug-n-play operation.

In commercial deployments, networking is not that straightforward. Think about it. How many computers are there in a broadcast center? Hundreds? Thousands? What about embedded systems, such as those in graphics playout devices, or nodes in storage and archive systems? A thousand doesn't sound like enough, especially if you add in all the front and back office systems.

Design, configuration, testing, and commissioning large networks is a difficult undertaking. Acquisition of the skills necessary to attain mastery of networking technology is a career-long pursuit. New technology appears daily.

Even though mastery of networking techniques, standards, design, and real-world implementation is a lifelong effort, every technologist in the media business should have some basic knowledge of networking. At the very least, it will allow intelligent communication of system concepts and detail among all engineering and support team members with system integrators and equipment vendors.

LET THERE BE CONNECTIVITY

The space race played a seminal role in the development of networking methodologies. In a state of shock after the Soviet launch of Sputnik, DoD directive 5105.15 established the Advanced Research Projects Agency (ARPA) and was signed on

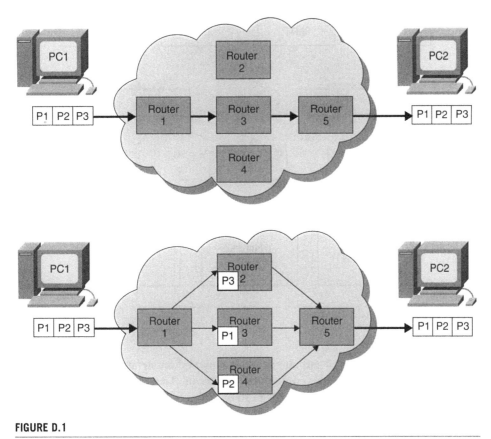

FIGURE D.1

Packet versus circuit switching

February 7, 1958 by President Eisenhower. Establishing network capabilities for defense systems that would survive a nuclear attack was a top priority.

The Semi Automatic Ground Environment (SAGE), an early network program, had created a countrywide radar system, and ARPA created the Information Processing Technology Office (IPTO) to pursue network R&D. J.C.R. Licklider was selected to head the IPTO and recruited Lawrence Roberts to head a network implementation project. Roberts based the technology on the work of Paul Baran, who had written an exhaustive study for the U.S. Air Force, which recommended packet switching (as opposed to circuit switching) to make a network highly robust and survivable.

Figure D.1 illustrates the difference between circuit switching and packet switching. In the illustration above, a connection and dedicated route is established during session initiation, analogous to making a telephone connection. Data packets P1, P2, … travel over this circuit, from routing node to node, to their destination. When the data transfer is complete, the circuit is broken down.

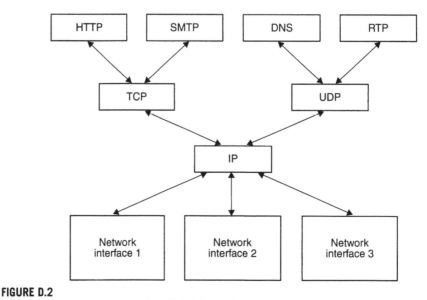

FIGURE D.2

The TCP/IP protocol suite

Now consider a national circuit-switched network under attack. If a portion is destroyed that is part of the connection route, all data communication ceases. Packet-switched networking solves this problem.

Packets contain the destination (and source) addresses. As they reach a router, the best next-hop router destination is determined and the packet is sent to this router. This process is repeated till the packet reaches its destination. The next packet follows the same process but may or may not take a completely different route.

In packet switching, if a portion of a network is destroyed, packets will be routed over a different path to their destination.

The first two nodes of what would become the ARPANET were interconnected between UCLA and SRI International in Menlo Park, California, on October 29, 1969.

Geographically dispersed networking will be discussed in greater detail in Chapter 7. The important concept discussed here is the difference between packet switching and circuit switching methodologies.

TCP/IP

A fundamental methodology that was developed and is relevant today to networks of all sizes is the TCP/IP stack. A stack is a representation of how protocols interrelate. TCP/IP is the enabling technology for the Internet and countless other network deployments.

TCP/IP utilizes the concept of four layers as shown in Figure D.2. These are the physical, internetwork, transport, and application layers.

aaa.bbb.ccc.ddd 4 octed dotted decimal IP address: each octet can range for 0–255

Class	First octet	Range	Net ID	Host ID	Number of nets	Number of hosts
A	0XXXXXXX	0 – 12	a	b.c.d	128 = (27)	16,777,214 = (224 - 2)
B	10XXXXXX	128 – 191	a.b	c.d	16,384 = (214)	65,534 = (216 - 2)
C	110XXXXX	192 – 223	a.b.c	d	2,097,152 = (221)	254 = (28 - 2)

For Network addresses the host address is always 0: 10.0.0, 128.bbb.255, 192.bbb.ccc.0
Broadcast addresses always have 255 for the Host octet: 10.0.0.255, 128.bbb.255.255, 192.bbb.ccc.255

IANA reserved private network ranges	Start of range	End of range	Total addresses
24-bit block (/8 prefix, 1 × A)	10.0.0.0	10.255.255.255	16,777,216
20-bit block (/12 prefix, 16 × B)	172.16.0.0	172.31.255.255	1,048,576
16-bit block (/16 prefix, 256 × C)	192.168.0.0	192.168.255.255	65,536

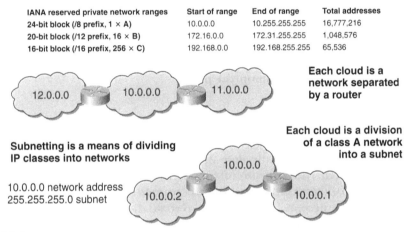

Each cloud is a network separated by a router

Subnetting is a means of dividing IP classes into networks

Each cloud is a division of a class A network into a subnet

10.0.0.0 network address
255.255.255.0 subnet

FIGURE D.3

IP addressing

The physical layer comprises the hardware: wires, line drivers, receiver, etc. The idea is to enable the higher protocols to run on any type of physical network deployment. Besides the physical portion, this layer has a logical component. This includes a unique media access control (MAC) number that consists of the manufacturer's assigned ID code and a unique number that creates an "address" of 6 bytes. Low-level, broadcast, and circuit-based transmission happens on this layer.

IP-directed packet routing occurs on the internetwork layer. An IP address has network and host portions as shown in Figure D.3. A thorough explanation of IP is beyond the scope of this book. However, examples of networking that follow will add details about how the address structure is used in network design.

Packet-switching networks require transmission control for reasons that will be described shortly. Packets can arrive out of order or not at all. The TCP layer uses mechanisms to detect and correct packet-transmission problems.

Programs that access the network are found on the application layer. TCP/IP specified four protocols:

- Hypertext Transmission Protocol
- Simple Mail Transfer Protocol

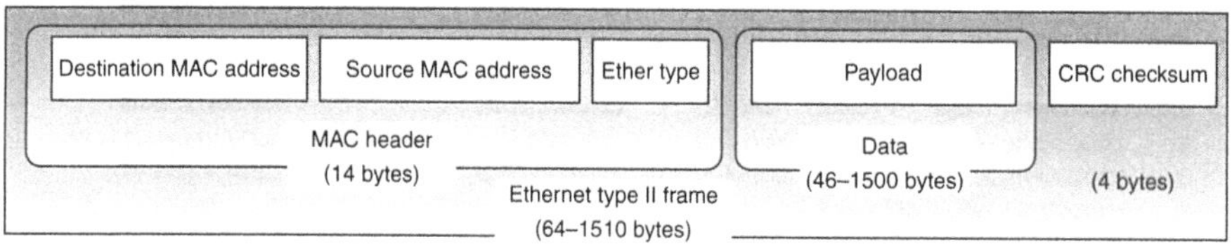

FIGURE D.4

Ethernet frame

- Domain Naming Service
- Real-Time Protocol

They will be described at appropriate points.

Ethernet

Many different protocols have been developed that operate on the physical layer. However, Ethernet, developed by Bob Metcalfe at Xerox in 1975, is probably the most widely utilized. The structure of an Ethernet frame is quite simple and illustrated in Figure D.4. Besides the payload and error checking, destination and source MAC addresses are present. These are used to route the packet to its destination as will be described shortly.

The development of Ethernet grew out of RF-based systems. A method was necessary to avoid collisions of transmissions on the same frequency. A method was implemented that checked to see if the frequency was in use before transmitting.

In the Ether

The term ÆTHER originated in the late 1800s. It was the medium that light waves were theorized to travel in. By the 1970s, it was considered to be nonexistent.

Shared-cable, carrier sense multiple access (CSMA) Ethernet has disappeared. When Ethernet is spoken of today, it means switched Ethernet. It is used to directly connect equipment with a routing device, a node to a node; no other packets will share the physical medium. Hence an Ethernet connect will attain data rates that are close to wire speed, the highest speed a medium can theoretically attain.

The payload can be up to 1,500 octets (eight serial bits) for a standard packet. Media networks may use jumbo frames, of up to 9,500 octets, to reduce header overhead and increase the data Transfer rate.

Internet protocol

IP packets contain two parts. The payload can be up to 1,500 bytes for a standard packet. A header, located before the payload, contains destination and source IP

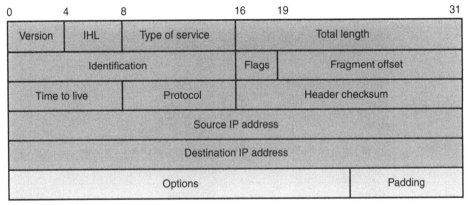

FIGURE D.5

Internet protocol header

addresses along with other information. Figure D.5 is an illustration of the fields in an IP packet header.

The header field functions are:

- Version—4 for IPv4; 6 for IPv6
- IP Header Length—number of 32-bit words forming the header, usually five
- Type of Service (ToS)/Differentiated Services Code Point (DSCP)—indicates QoS requirement of the network;
- Size of Datagram—the total length in bytes of the header and the data
- Identification—16-bit number that uniquely identifies the packet
- Flags—control packet fragmentation
- Fragmentation Offset—expressed in eight-byte blocks; offset from the start of the original packet before it is fragmented
- Time-To-Live—number of router hops allowed
- Protocol—service access point (SAP); indicates the type of transport packet being carried (e.g. 1 = ICMP; 2 = IGMP; 6 = TCP; 17 = UDP).
- Header Checksum—detects packet errors; packets with an invalid checksum are discarded by all nodes in an IP network
- Source Address—IP address of the original sender of the packet
- Destination Address—IP address of the final destination of the packet
- Options—indicates the presence of additional header fields

The maximum size of an IPv4 payload is 65,535 bytes.

Transmission

The transmission layer uses two protocols. TCP, as its name implies, is a mechanism that manages the transmission of data packets. It includes a handshake mechanism to ensure packet delivery and to initiate retransmission of lost packets.

Table D.1 The Well-Known TCP/IP Port Numbers

Service	Port	Function
HTTP	80	Web
HTTPS	443	Web (secure)
FTP	20,21	File transfer
FTPS	989,990	File transfer (secure)
Telnet	23	Remote login
SSH	22	Remote login (secure)
DNS	53	Host naming
SMTP	25	Internet mail
POP3	110	Client access
IMAP	143	Client access
NNTP	119	Usenet newsgroups
NNTPS	563	Usenet newsgroups (secure)
IRC	194	Chat
NTP	123	Network time
SNMP	161,162	Network management
CMIP	163,164	Network management
Syslog	514	Event logging
Kerberos	88	Authentication
NetBIOS	137–139	DOS/Windows naming

User datagram protocol (UDP) is a "best-effort" transmission method in which a transmitting device sends a packet without a handshake. As a result, there is no way to determine if the packet has been received at its desired destination.

Both these protocols are discussed in greater detail in Chapter 7.

Application ports

Many different applications can travel over the same network connection. The TCP/IP protocol uses a method that associates an application with a port. These can be a permanent port number assigned by the Internet Assigned Numbers Authority (IANA) to an application or a variable port number assigned during the connection. Table D.1 lists the "well-known" port numbers and their associated services.

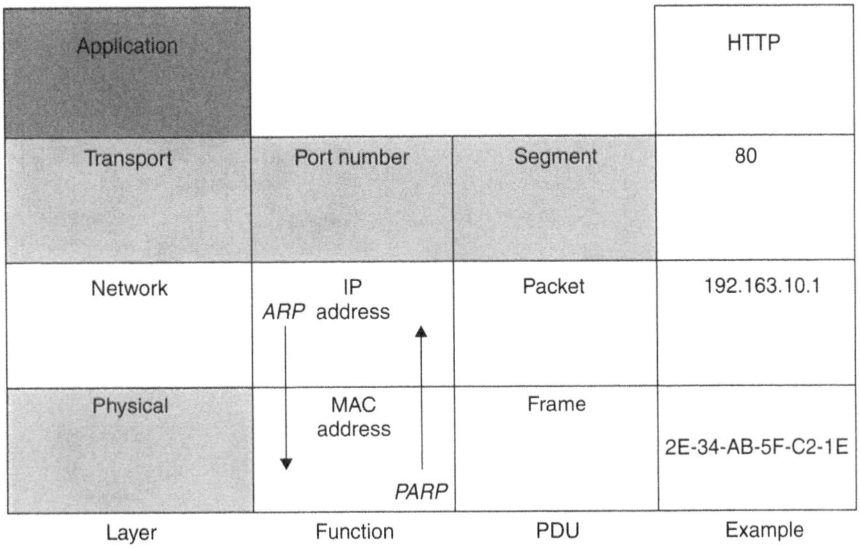

Layer	Function	PDU	Example
Application			HTTP
Transport	Port number	Segment	80
Network	IP *ARP* address	Packet	192.163.10.1
Physical	MAC address *PARP*	Frame	2E-34-AB-5F-C2-1E

FIGURE D.6

TCP/IP protocol stack details: Layer name, function, protocol data unit, and example

The well-known ports range from 0 through 1023. The registered ports are those from 1024 through 49151. The dynamic and/or private ports are those from 49152 through 65535.

UNIX OSs create a socket by binding a port, a protocol (TCP or UDP), and an IP address.

Network data transfer mechanics

Figure D.6 summarizes the relationship between layers in the TCP/IP protocol stack.

Details of exactly how a packet is transferred will be discussed in later chapters.

NETWORK TOPOLOGIES

Network configurations are implemented with a two-view perspective. There is the physical view that shows the way in which the devices are connected. This is what you will see on a system cabling diagram or detailed schematic drawing. The way in which IP addresses are structured can create a variety of logical network configurations. Good network design will take both into account.

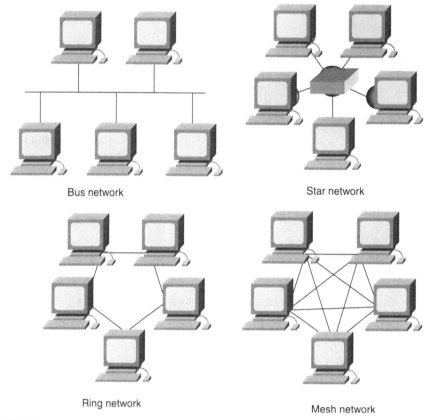

Bus network

Star network

Ring network

Mesh network

FIGURE D.7

Basic network topologies: star/hub, bus, ring, and mesh

Network configurations

There are a handful of network topologies that can be configured by using a combination of logical and physical design techniques. As illustrated in Figure D.7, the basic topologies are bus, hub (star), ring, and mesh.

The visual representations are fairly self-explanatory. Devices in a network are frequently referred to as nodes and can be computers, storage devices, printers, or any other device that is connected to another device.

The OSI model

The TCP/IP protocol stack is an abstract representation of computer network interconnections. There are other approaches. The open system interconnect (OSI) is another model that is used to describe networks.

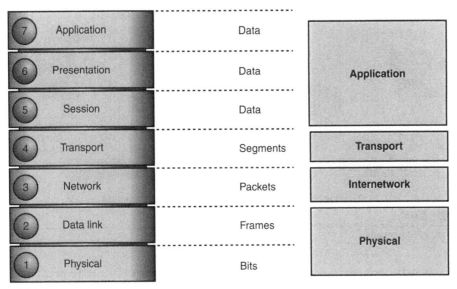

FIGURE D.8

The OSI seven layer model compared to the TCP/IP model

The OSI conceptual protocol stack comprises seven layers. As can be seen in Figure D.8, there is a direct metaphor with the TCP/IP model. Both conceptually describe identical systems.

Powerful as the OSI model is, it has never really caught on in network design and implementation; TCP/IP remains the dominant protocol stack. However, the seven OSI layers are used to discuss the relationship of networked computer systems. For example, the physical layer in TCP/IP includes the logical MAC address. In the OSI model, this is the data link layer (layer 2).

Network operating system

Routers and switches are computers optimized for network applications. An OS, network operating system (NOS), or internetworking operating systems (IOS) enables implementation of routing and switching as well as other sophisticated traffic engineering and security functions.

Routing devices

Routing devices fall into a number of categories based on functionality. The OSI provides a more detailed granularity than the TCP/IP model. OSI layers are used in the IT industry to describe network device operation.

The simplest network interconnect device that does not perform any routing operations is the hub. A hub connects all devices together in a bus configuration and sends and receives all packets from all devices. Hence, a hub is a bus topology from both the physical and logical viewpoints. It can be considered a layer 1 device.

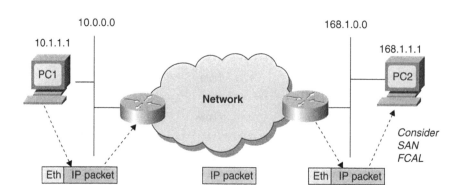

IP packet contains the destination and source IP addresses

FIGURE D.9

Packet encapsulation

A switch performs connection-oriented routing and operates at layer 2. Just as with a phone call, a session must be set up and then torn down. The paths between devices are dedicated connections, and packets always follow the same path as shown in Figure D.8.

A router is a layer 3 device and, as its name implies, routes packets to their destination address. This can be direct or over a number of "hops." A "hop" occurs when a packet passes through a router. Routers must decode the address and make decisions about which path to send the packet on to its destination based on routing protocol and routing metrics. This is the packet switching method shown in Figure D.1.

TCP is a layer 4 protocol and is discussed in Chapter 7.

Packet encapsulation and PDUs

As data travels across a network, it undergoes a series of operations. As packets travel down the protocol stack, they are wrapped in the next layer protocol; as they move up the stack, they are unwrapped. This process is called encapsulation and is illustrated in Figure D.9.

A packet originates from a device with an IP address of 10.1.1.1 as an IP packet. It is transmitted as an Ethernet packet. This is accomplished by encapsulating the entire IP packet in the payload of the Ethernet frame. When the Ethernet frame enters the frame relay network, the IP packet is unwrapped and traverses the network as an IP packet.

The IP packets travel from source to destination on a network in undefined, connectionless paths. Routers determine the best path for any given IP packet to take.

When the IP packet arrives at its destination, it is wrapped as the payload in an Ethernet packet. It is now compatible with the destination network protocol.

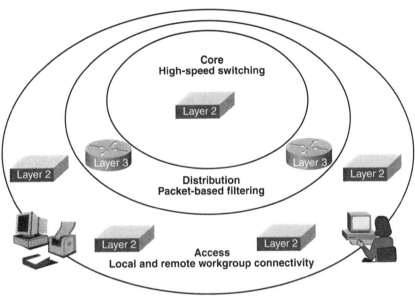

FIGURE D.10

Core, aggregate, and access (edge) network model

Three layers

A three-layer model has evolved by combining basic network topologies. Figure D.10 depicts the architecture.

Beginning at the edge or access layer, computer resources, classified by work-groups, function, or location, are connected to OSI layer 2 switches. Each of these connections operates at the full bandwidth over a dedicated connection. These edge switches are connected to the aggregate layer routers. To ensure system reliability, edge switches are connected to more than one aggregate router. Uplink data rates are usually an order of magnitude higher than port speeds.

Connectionless packet routing at layer 3 routers performs rule-based filtering. This is where ACLs are applied.

In the core aggregate, routers are connected over a mesh topology. High-speed switching is used to transfer data packets.

Domain naming system

The domain naming system (DNS) resolves human readable names, such as www.philipcianci.com, into IP addresses. It is much easier to connect to a device using a name rather than a long number. Figure D.11 describes DNS and address resolution protocol (ARP), two methods used to route packets to their proper location.

The DNS is part of the core TCP/IP protocol stack. Its purpose is to resolve a name to an IP address. PC1 sends a request for an IP address for the computer

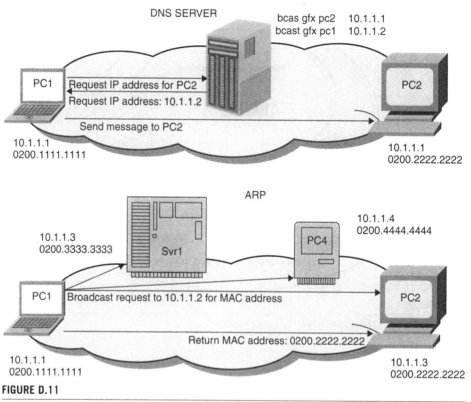

FIGURE D.11

DNS, ARP, MAC, and IP addresses

named PC2 over the network to the known address of a DNS server. The DNS server replies with the IP address and the message is sent to PC2.

In a layer 2, switched environment, packet routing is based on MAC addresses. Therefore, a method is needed that associates IP and MAC addresses. ARP accomplishes this task. PC1 broadcasts a request to every device on the network announcing it is looking for MAC address of PC2, which it knows is IP address 10.1.1.2. PC2 has this IP address and answers by sending its MAC address back to PC1. PC1 can now use the MAC address to construct an Ethernet frame.

Appendix E: Image processing

In order to process an image on a computer system, the image must be converted to digital form. PCM, which we discussed earlier, is most frequently used for audio, but it can be applied to conversion of an image from analog to digital. However, the process is more complicated than it is for audio.

The first step for an imaging device is to convert photons of light to an electrical potential. Then the image is scanned, in a method that is the same as when the image is eventually displayed. Scanning results in a series of pixels.

Here's how it works, in rudimentary terms. A pixel has characteristics that can be described in terms of its red, green, and blue components. Each of these has an intensity value, which maps to a code word. The pulse is the temporal relationship of pixels, and each pixel represents a sample. Again, modulation is a result of varying the amplitude of the sampling clock.

Installation of a circuit board with video input and output in a computer system is necessary to capture and digitize video. These "frame grabbers" were the first steps in the emergence of digital imaging; they converted frames of video to files of data.

DIGITAL PHOTOGRAPHY

Chemical-based film photography has all but disappeared. Film has superior resolution compared to digital image formats; in fact, a driving requirement of HDTV development was the image resolution approach of film. Yet even with film's high-quality imagery, convenience has won out and consumers have flocked to digital cameras and computer storage rather than negatives and video displays. Digital picture frames abound on office desks.

Ten megapixel digital cameras are the choice of many consumers. A PC is an image storage system and processing lab. Got red eyes on the people in your photo? Don't worry; the software will get rid of it for you. Today's image processing software is very sophisticated, and the latest PCs can run the programs with ease. Twenty years ago, those in digital imaging would have killed to have the kind of processing power most people now have on their desktops.

But 10 megapixels take up a lot of hard drive space. In light of the expense and scarcity of disk space, imaging research looked for ways to reduce the size of digital image files. The result is JPEG, which bears the name of the ISO working group that standardized the technique: the Joint Photographic Experts Group. The formal standard was adopted in 1994 as ISO 10918-1.

The first digital cameras for the consumer market that worked with a home computer were the Apple QuickTake 100 camera (1994), the Kodak DC40 camera (early 1995), the Casio QV-11 (with LCD monitor, late 1995), and Sony's Cyber-shot Digital Still Camera (1996).

The family album has given way to the ease of electronic storage. Collection of your images on your PC is automated. Printers produce high-quality copies, and you don't even need a PC to print: just plug the photo memory card into the printer and you're all set.

Digital cameras are everywhere. If you've got one in your cell phone or PDA, then just aim, shoot, and e-mail; you'll never miss capturing and sharing magic moments.

Of course the television had to get into the act. All you have to do is select the pictures you want and let the software burn a DVD or CD that you can play back with your component DVD player. The digital camera, PC, and television are now part of the home network.

DIGITAL CINEMA

Just as motion picture technology was born from still photography, digital still images are the basis for digital cinema. Frame rates and pixel grids determine the amount of detail present and how much of that detail survives frame to frame.

Digital production, distribution, and presentation have advantages that the film industry cannot overlook. Many cinematographers still prefer the "film look" to video, but movie production has gone digital using a process dubbed Digital Intermediary (DI).

Compare special effects of 20 years ago with those of today. If the desired effect could actually be achieved, it took an army of artists, working film frame by frame to produce a clip, a very expensive and time-consuming process. In a DI workflow, the first step is converting the film content, frame by frame, to digital video. This is done by a datacine machine, the descendent of a telecine machine.

Once in digital form, software-based editing and image processing systems can do anything—yes, anything—with the image. Although the conversion of film-to-digital video process can be expensive and time-consuming, the time and money saved during production and the freedom from artistic limitations make the overall DI process cost-effective and creatively liberating.

Copying a film master to produce thousands of distribution copies is another expense that can be reduced with films in digital format. Movies can be transferred as digital files over commercial content distribution networks to digital theaters. By comparison, the network connection is cheaper than producing and delivering copies to hundreds or thousands of theaters.

Digital cinema is still in its infancy. But the number of locations where digital projection systems have been installed is growing rapidly around the world. A major expense, in addition to the initial installation of a digital projection system, is having sufficient disk space to store the digital "film." We're talking about terabytes. Consider a 10-screen multiplex cinema with a movie for each screen. Don't forget the space needed for a safety, backup copy.

Presentation of digital movies, as with DTV, is perfect. Digital files don't wear out, so there aren't any film scratches or other deterioration over viewing lifetime. The 1,000th presentation is at the same quality as the first.

Besides the scale of the presentation, digital cinema has an advantage over television. One is that digital movie image resolution starts at 2K, roughly the same as 1080p24, and then increases to 4K. Details not visible in HD can be seen in 4K digital cinema.

DIGITAL SIGNAGE

Fremont Street in downtown Las Vegas has long been a kaleidoscope of visual images since the first casinos installed neon lights. But they pale in comparison to the domed display, close to a 1/4-mile long and 100-feet wide, and the "light show" that's now presented nightly. The original system, which began construction in 1994, consisted of dozens of custom computer circuit boards controlled by a custom authored software application. It was upgraded to LED technology and higher image resolution in 2004. The resulting experience is unforgettable.

Conspicuously in sports arenas and shopping malls, static billboards are being replaced with LCD displays. Coupled with computer control, sophisticated sequences of video, graphics, and text can be programmed and run with ease.

The technology is called digital signage and is an emerging outlet for all types of media. It is expected to be a large player in the future of out-of-home media.

IMAGING CHARACTERISTICS

The emergence of the PC, video games, and Internet was fueled by the development of the general-purpose microprocessor. Rather than being large electronic computer systems designed for a specific purpose, the Z80, Motorola 6800, and Intel 8080 were discrete devices that could perform a variety of mathematical, logical, data movement and storage functions.

Math coprocessors were tweaked for number crunching and supplemented the core microprocessor, and became an option for PC systems in the 1980s. An Intel 80286 would be supplemented by an 80287 math coprocessor. Next down the R&D pipeline were digital signal processors (DSPs). These were specifically designed for the heavy computational demands and data transfer speeds necessary for media applications.

A generic image processing workflow is presented in Figure E.1. It is very simple: capture the image, store the image, process the image, store the results, and then display the image.

Storage is a key component of digital media processing systems. Only recently have the prices of RAM and disk storage reached a cost-effective level for general production use. Remember that half-million-dollar SGI Origin I talked about earlier? In 2000, 100 GB of RAID (Redundant Array of Inexpensive Disks) cost $125,000.

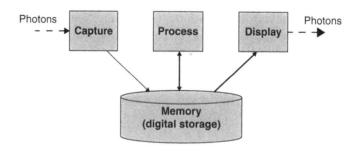

FIGURE E.1

Digital image processing pipeline

A dual-port HD-SDI I/O card was $20,000. Not very many broadcasters could afford this, except CBS, which used the system for its first HD broadcasts.

By the turn of the millennium, the convergence of the DTV and Internet revolutions was becoming feasible if not quite yet a reality. About 5 years later, handhelds attain similar capabilities. All this has been enabled because processing power, storage and memory read/write speeds, and internal bus bandwidth have reached the ability to support real-time audio, video, and graphics.

Digital image formats can be broken down into their constituent building blocks. These are aspect ratio, pixel grid, frame rate, scanning method, and color space. Each combination of these attributes can be considered a presentation format. Let's take a look at each.

Aspect ratio

Probably the first thing anyone noticed about HDTV was the different screen geometry. The 16×9 aspect ratio looked a lot larger than 4×3 analog TV. In a multiplatform universe there are other sizes to reckon with. Film uses ratios up to 2.25×1, and a 6×4 photo is 1.5×1. Aspect ratio has now become a concern for content repurposing.

Preserving the artist's visual intent can be difficult. Consider the scene in Figure E.2 presented in four different aspect ratios. We'll consider the 4×3 image the source; this is the way the camera operator intended the shot to be framed. Converting to 6×4, some of the sky is lost, but there is not much of a dramatic impact on the scene. Cropping for 16×9 obviously changes the image aesthetics. The subjects seem to be integrated with the falls. In the 2.25 crop the focus has clearly shifted to the subjects and the falls are more of a backdrop.

Pixel grid

The concept of breaking an image up into a pixel grid is characteristic of digital imaging. Figure E.3 shows the steps in the progression. Film is an emulsion; detail resolution depends on the fineness of the particles. Film is generally considered the highest attainable resolution.

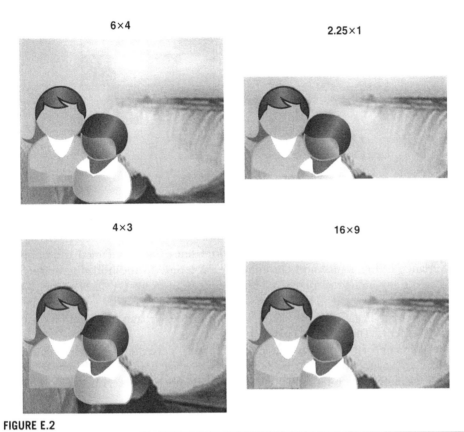

FIGURE E.2

Comparison of imaging aspect ratios

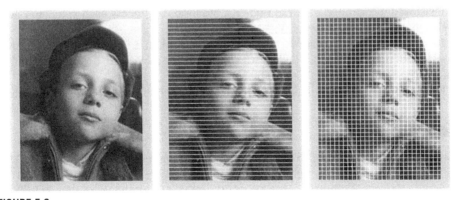

FIGURE E.3

Image capture and display techniques

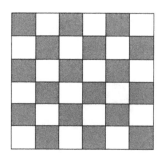 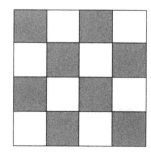

FIGURE E.4

A comparison of pixel grid resolution. The grid on the left has nine times the resolution of the one on the right

Analog TV dissected the image into distinct lines but maintained the continuous variance of luminance and chrominance on a line. In the horizontal direction, there aren't any individual pixels. CTR technology, even for DTV, still employs this mechanism.

But when an image is digitized, the lines are further broken down into individual picture elements. The result is a pixel grid.

The number of vertical and horizontal pixels determines spatial resolution. Consider an image that is 400 by 600 pixels. (Please note the numbers selected in these examples are for simplicity.) The total number of pixels is 240,000. On a 4 × 6 inch print, this equates to 100 pixels per inch.

Now consider an image with three times the vertical and horizontal resolution. 1200 by 1800 results in 2,160,000 pixels at 300 pixels per inch, three times the number in both the horizontal and vertical directions: an overall increase by nine times over the same area. Figure E.4 compares two pixel grids. Note the differences in detail capture capability of each pixel grid.

Pixel grids are naturally inclined to digital technology. With red, green, and blue values stored in memory, a simple two-variable loop can access each value and activate each element (see Listing E.1).

To process an image, data is read out in this fashion from a frame buffer, processed, and then stored. To display the image, it is moved to the display buffer, which, when fully loaded, is sent to the screen.

The number of times the display is updated in a second is called the refresh rate.

Refresh rate

Motion in motion pictures is based on the visual attribute of image persistence. The physiology and neurology of vision retains an image for a short period of time. This visual hysteresis is exploited to create the illusion of continuous motion. Research has revealed that if an image is repeated at a rate of over 24 times per second, jitter is not visible to the human eye.

A small variety of refresh rates are used across the globe. In the United Kingdom and Europe, 25 and 50 Hz are used. The selection has to do with meeting the perceptual hysteresis minimum because electrical AC power is at 50 Hz. By using identical frequencies, techniques can ensure that artifacts produced by differences in refresh and power rates aren't noticeable.

The United States uses 60 Hz AC power, so the original NSTC I black-and-white TV system was at a 60 Hz display refresh rate. However, the development of color television necessitated the need to slightly alter the refresh rate to 29.97 to avoid audio carrier issues. The ratio between the two is 1000/1001.

But there are fundamental issues when it comes to TV and PC displays. Television in the United States is based on a 29.97 Hz frame refresh rate, a lingering vestige of the analog NSTC II system. PCs operate at 60 Hz and other "exotic" rates.

The other difference is the scanning method.

Scanning method

As just discussed, television imaging dissects an image into a series of horizontal lines. The obvious way to do this is to start at the top of the image, scan the line, jump down the distance of one line and move back to the beginning, then scan the next line, and so on until the entire frame has been processed. This sequential scanning of line 1, then line 2, is called progressive scanning. Progressive scanning is exclusively used in computer displays. Refresh rates are in terms of frames per second.

Early TV systems were constrained by the bandwidth of the terrestrial delivery channel. In the United States, this was, and still is, 6 MHz; in PAL systems deployed in

```
Display_Frame
{
for y = 1 to Vertical_Resolution
    {
    for x = 1 to Horizontal_Resolution
        {
        get (x, y, Red, Green, Blue);
        display (x, y, Red, Green, Blue);
        x = x + 1;
        }
    y = y + 1;
    x = 1;
    }
y = 1;
x = 1;
}
```

LISTING E.1

Pseudo C code pixel grid frame display algorithm

the United Kingdom and Europe, channel bandwidth is 8 MHz. This required some form of data reduction in order to fit the TV signal into the channel space.

A technique where a frame of video lines is divided into two fields, one of odd lines and the other of even, was devised. The technique, known as interlacing, is used today in DTV systems. Any format with an "I" in it denotes interlaced scanning.

DTV offers the choice between delivering content using either progressive or interlaced scanning. Conversion between the two is not trivial and is discussed in Chapter 9.

Color space and gamut

Accurate rendition of color is imperative for photography. The driving motive of photography is realism. Regardless of whether chemical or electronic image reproduction technology is used, color fidelity is dependent on the spectral properties of each of the primary colors. Images are "colorized" by the use of three-color systems. Printing to paper is a color subtractive process and uses cyan, yellow, and magenta, along with black. Video displays use red, green, and blue primaries in an additive process.

All color primaries are not equivalent. They are dependent on the chemical properties of the material used to reflect or produce light. This causes differences between primary colors used in negatives, paper, imager, and displays. As shown in Figure E.5, the location of red, green, and blue primaries on the CIE color chart varies. If the R, G, and B values are connected to form a triangle, it becomes obvious that different sets of primaries produce different ranges of color.

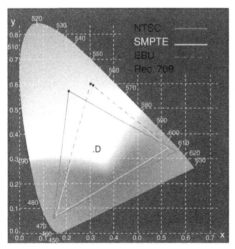 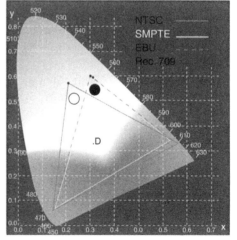

FIGURE E.5

Color primaries and gamut mapped on the CIE color chart

The area inside the triangle is called the color gamut. All colors in the natural world are beyond any set of primaries that can be reproduced on a display system. It is obvious that some sets cannot produce some colors that other sets can.

But there is a more subtle effect of color primary variation. Consider adding equal amounts of R, G, and B to make white. The point where the primaries intersect to form white is in a different place for each set of primaries. Amazingly all whites are not the same white. The whiteness of a white is expressed as color temperature: 6500 K for daylight. Lower values tend to be more orange; higher, bluer.

While computer systems work with red, green, and blue primaries, TV systems exploit human visual system characteristics to reduce the amount of information so that the highest quality image can fit in the transmission channel.

Red, green, and blue image information is converted into luminance and chrominance signals. Figure E.6 shows the waveforms for the color bar test signal.

As the equations show, different amounts of red, green, and blue are combined to produce a luminance signal "Y" and two color difference signals "B-Y" and "R-Y." The conversion equations are based on the sensitivity of the eye to red, green, and blue.

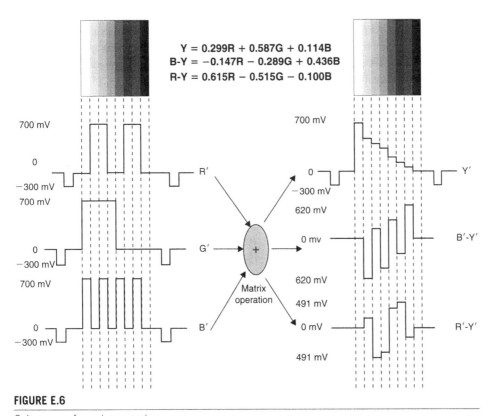

$$Y = 0.299R + 0.587G + 0.114B$$
$$B{-}Y = -0.147R - 0.289G + 0.436B$$
$$R{-}Y = 0.615R - 0.515G - 0.100B$$

FIGURE E.6

Color space format conversion

Bit depth and data volume

Each pixel is represented by a red, green, and blue value. If each of these is a byte (8 bits), the total number of bytes to store each of the image examples is 5,760,000 for the 400 × 600 image and 51,840,000 for the 1200 × 1800 image. The resolution of each color is 256 distinct values, and the total number of colors that can be produced is 256 × 256 × 256 or 16,777,216.

Professional production will use 10-, 12-, or even 24-bit color depth; 8 bits are sufficient for consumer applications. The longer words are necessary to avoid artifacts created by accumulated round-off errors during image processing.

Color sampling

Another technique that exploits human visual characteristics is the fact that the eye is less sensitive to color detail than it is to luminance detail. Hence color information can be reduced.

When red, green, and blue are converted to Y, there is a one-to-one correlation; a triplet of RGB produces one Y pixel. But with the color difference signals, the conversion is not one-to-one. Depending on the color sampling technique, one pair of chrominance pixels may be produced for every other RGB triplet on a line. This cuts the amount of chrominance data into half. A further reduction can be achieved by only producing chrominance data for every other line, resulting in one pair of color data points for four luminance pixels.

COMPRESSION IN IMAGING

Researchers realized that development of techniques to reduce the amount of data in a digital image would reduce the amount of storage and cost necessary to build digital image libraries. Business people understood that this would help bring about mass commercialization and consumer uptake of digital photography.

Another driver of compression technology development was the need to compress images for digital high-definition television. Audio compression was developed to introduce surround sound to movies. Internet delivery also played a role.

TIFF was an early compression format that is still in use today in image archives. TIFF is a lossless image compression technique. Lossless image compression uses data reduction techniques that will exactly reproduce the original image data; byte for byte, the data is identical to the original.

Lossy, or perceptual, compression removes image detail that is imperceptible to the human eye. The reverse process, decompression or decoding, results in a less than perfect reconstruction of the original image; byte for byte, the data may differ. The trick is that people won't perceive the visual information removed as missing in the reconstructed image. Similar psychoacoustic attributes are used in audio compression.

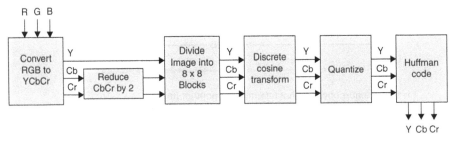

FIGURE E.7

JPEG processing block diagram

JPEG

The Joint Photographic Experts Group, an imaging industry consortium, was formed in 1986 to consolidate investigations into image compression and to develop a consensus standard. The resultant technique has become an acronym known to all: JPEG.

JPEG is a lossy image compression format. Although a JPEG file can be encoded in various ways, most commonly it is done with JFIF encoding. The encoding process consists of several steps as shown in Figure E.7.

- The color space of the image is converted from RGB to YCbCr (Cb and Cr are forms of B-Y and R-Y). This step is optional.

- Chroma data is reduced, usually by a factor of 2.

- The image is split into blocks of 8 × 8 pixels; each block of Y, Cb, and Cr data is converted from the spatial to the frequency domain via a discrete cosine transform (DCT).

- The amplitudes of the frequency components are quantized. Resolution is determined by how each of the frequency components is quantized. The more the bits used for high-frequency information, the higher the image resolution.

- The resulting data volume is reduced with a lossless Huffman encoding algorithm.

The decoding process reverses these steps.

MPEG-1 and DVDs

CDs had been such a business success that it was only natural to envision a way to store movies on a similar optical medium. The vision of potential profits by heavy hitters in the consumer electronics industry launched a competition reminiscent of the VHS/Beta battle.

In the early 1990s, two high-density optical storage standards were being developed: MultiMedia Compact Disc (MMCD), by Philips and Sony, and the Super Density

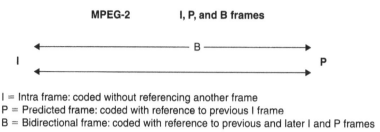

FIGURE E.8

The relationship among I, P, and B frames

Disc (SD), supported by Toshiba, Time-Warner, Matsushita Electric, Hitachi, Mitsubishi Electric, Pioneer, Thomson, and JVC. Legend has it that IBM President Lou Gerstner led an effort to unite the competitors behind a single standard.

Eventually Philips and Sony abandoned their MMCD format and agreed upon Toshiba's SD format with two modifications related to the servo tracking technology. This enabled the inclusion of Philips and Sony technology in the standard and the derivation of revenue from intellectual property IP licensing.

Video had to be digitized; optical recording is a digital technique, storing ones and zeros in the glass substrate. To resolve the obstacle, the JPEG tool kit added features that supported motion. The result was MPEG-1 (the M is for motion). Video storage on DVDs initially used MPEG-1 and has moved on to MPEG-2 compression.

The first DVD players and discs became available in 1996 in Japan, March 1997 in the United States, 1998 in Europe, and 1999 in Australia. Video resolution was limited to 352 × 240 at 30 fps. This produces video quality slightly below the quality of conventional VCR videos.

Perhaps the most innovative feature of MPEG-1 was the technique of encoding some images as only the difference between one image and a previous image. In practice, three frames types are used: I, P, and B.

- I—Intraframe: an anchor frame that can be decoded without reference to another frame; in essence, a JPEG image.

- P—Predicted: a frame that encodes the difference in an image with respect to a previous I frame.

- B—Bidirectional: a frame that encodes the difference between this frame and earlier of later I or P frames.

Figure E.8 shows the relationship between the types of compressed frames.

Further developments to the original MPEG-1 standard resulted in MPEG-2 and MPEG-4 standards. Ongoing work by MPEG working groups is the development of MPEG-7 and MPEG-21, both of which address content management.

Index

331